T0185606

SpringerBriefs present concise summaries of cutting-edge research and practical applications across a wide spectrum of fields. Featuring compact volumes of 50 to 125 pages, the series covers a range of content from professional to academic.

Typical publications can be:

- A timely report of state-of-the art methods
- An introduction to or a manual for the application of mathematical or computer techniques
- A bridge between new research results, as published in journal articles
- A snapshot of a hot or emerging topic
- An in-depth case study
- A presentation of core concepts that students must understand in order to make independent contributions

SpringerBriefs are characterized by fast, global electronic dissemination, standard publishing contracts, standardized manuscript preparation and formatting guidelines, and expedited production schedules.

On the one hand, **SpringerBriefs in Applied Sciences and Technology** are devoted to the publication of fundamentals and applications within the different classical engineering disciplines as well as in interdisciplinary fields that recently emerged between these areas. On the other hand, as the boundary separating fundamental research and applied technology is more and more dissolving, this series is particularly open to trans-disciplinary topics between fundamental science and engineering.

Indexed by EI-Compendex, SCOPUS and Springerlink.

More information about this series at http://www.springer.com/series/8884

Tin-Chih Toly Chen

3D Printing and Ubiquitous Manufacturing

 Springer

Tin-Chih Toly Chen
Department of Industrial Engineering
and Management
National Chiao Tung University
Hsinchu, Taiwan

ISSN 2191-530X   ISSN 2191-5318   (electronic)
SpringerBriefs in Applied Sciences and Technology
ISBN 978-3-030-49149-9   ISBN 978-3-030-49150-5   (eBook)
https://doi.org/10.1007/978-3-030-49150-5

This Springer imprint is published by the registered company Springer Nature Switzerland AG
The registered company address is: Gewerbestrasse 11, 6330 Cham, Switzerland

Contents

Chapter 1
Introduction

1.1 Three-Dimensional Printing

Three-dimensional (3D) printing is a new approach for additive manufacturing, which is to create a 3D object by forming successive layers of materials by using a 3D printer controlled by a computer [1]. In this way, 3D printing can fabricate products or components with very complex shapes. Another advantage of 3D printing is the easiness of integrating with computer-aided design (CAD) and computer-aided manufacturing (CAM), so that a 3D object can be directly made from the model generated by a CAD/CAM software package. Naturally, existing CAD/CAM software vendors were pioneers in this field. For example, AutoCAD has supported reading a 3D model in the stereolithography (STL) file format and output an STL file to a 3D printer since AutoCAD 2011. Catia can also save a 3D model in the STL format [2]. Other acceptable 3D printing file formats include OBJ, VRML, PLY, and ZIP. 3D printer vendors, CAD/CAM software companies, 3D printing service providers, and 3D-printing communities have been maintaining online databases of 3D objects. A user can download a 3D model resembling his or her idea from these websites and then modify it. In addition, offline databases such as 3D anthropometry databases and medical databases (including computed tomography and magnetic resonance imaging databases) can also be converted to be printable, as illustrated in Fig. 1.1.

The development of 3D printing dates back to the early 1980s [3]. After decades of research and development, the application of 3D printing has transited from visualization and prototyping to mass customization and mass production [4–6]. American Society for Testing and Materials classified additive manufacturing processes into seven categories [7]:

- Vat photopolymerization,
- Material jetting (MJ),
- Binder jetting (BJ),

© The Author(s), under exclusive license to Springer Nature Switzerland AG 2020
T.-C. Chen, *3D Printing and Ubiquitous Manufacturing*,
SpringerBriefs in Applied Sciences and Technology,
https://doi.org/10.1007/978-3-030-49150-5_1

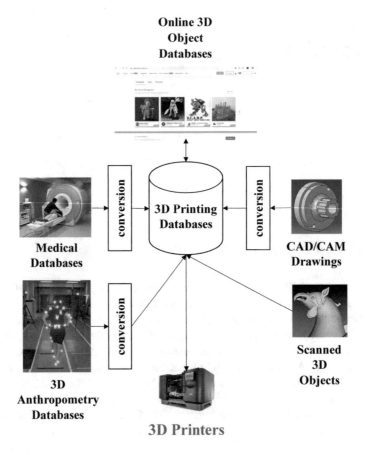

Fig. 1.1 Data sources for 3D printing

- Material extrusion,
- Powder bed fusion,
- Sheet lamination, and
- Directed energy deposition.

Currently, the most popular 3D printing technologies include fused deposition modeling (FDM), stereolithography (SLA), masked stereolithography (MSLA), digital light processing (DLP), selective laser sintering (SLS), direct metal laser sintering (DMLS), selective laser melting (SLM), electron beam melting (EBM), MJ, drop-on demand (DOD), and BJ. These 3D printing technologies differ in the types of raw materials used, forms of raw materials, and principles followed. The major types of materials for 3D printing include resin (or plastics), metal, gypsum, and sand. An introduction of various 3D printing technologies refers to All3DP [8].

Three-dimensional printing has been applied in numerous industries. For example, in the automotive manufacturing industry, automotive makers have applied 3D printing technologies to prototyping, design validation, small-scale mass production, and final inspection [9]. Every year, automotive makers apply 3D printing to prototype or manufacture more than 100,000 parts and/or molds [10]. Functional spare parts of vehicles, engines, and platforms are also tested using 3D printing [9]. In the medical and healthcare industry, 3D printing technologies have been widely applied to make personalized products such as hearing aids, artificial ears, prostheses, rehabilitation aids, orthopedic surgery guide plates, artificial joints, and dental implants [11]. Medical products made using metal printing have porous titanium structures, are more ergonomic, and have higher performance, far exceeding the limitations of traditional manufacturing processes. 3D-printed titanium parts or products are prevalent in various industries such as aerospace, chemical, and biomedical industries due to excellent properties including lightweight, high specific strength, high chemical resistance, and biocompatibility [12]. In the biomedical industry, titanium is used to make implants because of its load-bearing and biocompatibility, while in the aviation industry, the resistance to corrosion and lightweight of titanium is emphasized [13–15]. Titanium is about 45% lighter than steel [16]. The replacement of steel with titanium has reduced the weight of a 777 airplane by 5800 lb [17]. A lighter weight also reduces the required fuels and carbon emission, which further promoted the usage of titanium in aircraft. In addition, titanium is twice as strong as aluminum that is frequently used to build the overall structure of an airplane [16, 17]. In addition, a traditional manufacturing process may be relatively complex, have long cycle time, and be difficult to maintain precision. The application of 3D printing technologies has overcome these problems to reduce manufacturing costs and shorten cycle time [13]. For example, in the construction industry, 3D printing technologies have been applied to build simple and affordable models rapidly. The application of 3D printing technologies also motivates innovative ways of production. For example, architects proposed a direct method of construction by applying 3D printing technologies [18]. 3D printing, cyberphysical systems, and the Internet of things (IoT) have brought about the third industrial revolution [19].

Four technical challenges faced by 3D printing researchers and practitioners are time-consuming 3D object design, limited types of usable materials, low precision, and low productivity, as illustrated in Fig. 1.2 [2]. To address these challenges, the following treatments have been taken:

(1) A major research trend in this field is to increase the number of types or suitability of materials that support the printing of a specific 3D object [5].
(2) Another focus is on how to improve the quality of a printed 3D object [20, 21].
(3) 3D scanners that can be used to scan objects to produce 3D models have become more and more popular and affordable [22]. However, this does not work for 3D objects that do not physically exist.
(4) Ubiquitous manufacturing (UM) systems based on 3D printing have been established to enhance the productivity of mass customization using 3D printers [23].

Further, to further enhance the effectiveness and efficiency of 3D printing applications, the following managerial issues need to be addressed [2]:

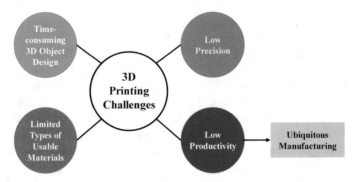

Fig. 1.2 3D printing challenges

- Three-dimensional object database management: Although most public 3D object databases do not allow or charge for commercial usage, it is better for a factory to confirm whether any similar 3D object exists in public databases before designing a new 3D model. In addition, 3D objects in different databases can be classified and aggregated, which requires some kind of artificial intelligence. In this way, the efficiency of searching for a specific type of 3D objects can be considerably enhanced.
- Intellectual property rights of 3D printing: A CAD file created by scanning a 3D object may not be copyrightable. In addition, it is easy to counterfeit a 3D object with its 3D model on hand by using a 3D printer, which imposes a great challenge on data security. Although the patent for an existing 3D printing technology may have expired, the patent for a product or its parts has not. Since 3D printing is more suitable for making smaller parts, it is most likely to infringe on the patents of such parts.
- Business innovation: Most hubs provide online catalogs of 3D models that have been collected for downloading or purchase. If a user purchases a 3D model, some hubs print and deliver the printed 3D object to the user. The related costs and expenses become the revenues of the hub. Alternatively, other hubs forward an order to the designer for processing. The hub receives a commission from the designer. These business models are based on e-commerce applications, including business-to-customer (B2C) e-commerce (with the hub as a business and a user as a customer) and customer-to-customer (C2C) e-commerce (between an uploading user and a downloading user).
- Ubiquitous manufacturing: It is easy to build up 3D printing capacity. As a result, it is easy to find many 3D printing facilities in a region to support UM.
- Lean manufacturing: Ubiquitous manufacturing based on 3D printing has reduced or even eliminated the assembly lines and supply chains of many products, making the whole industry leaner. In addition, with 3D printing, a product can be manufactured in a print-on-demand manner, eliminating the necessity of maintaining an inventory for the product, which conforms to the concepts of "pull systems" and "no inventory" in lean manufacturing.

- Globalization and deglobalization: With 3D printers, it is now possible to build manufacturing capacity at any place, which is a major step toward further globalization. By contrast, 3D printing destroys the industries in low-income countries that focus on production, because some products will no longer be produced by them but by the designer directly through 3D printing, which leads to deglobalization.
- Feasibility evaluation and optimization: When considering the feasibility of a 3D printing application, technical challenges are usually more critical than managerial concerns. However, when a 3D printing application is already in progress, managerial concerns should be addressed to optimize the 3D printing application.

1.2 Ubiquitous Manufacturing

Ubiquitous computing is a concept in software engineering and computer science meaning that computing can be made anytime and everywhere. Ubiquitous manufacturing is an application of ubiquitous computing in the manufacturing sector to enable convenient, on-demand network access to a shared pool of configurable manufacturing resources, including computer software and hardware, equipment, and manufacturing capability [23]. Many advanced manufacturing technologies, such as lean manufacturing, cloud manufacturing, manufacturing grid, global manufacturing, virtual manufacturing, agile manufacturing, Internet manufacturing, and additive manufacturing, have contributed to UM. Many UM studies equated UM with cloud manufacturing [24]. However, unlike cloud manufacturing, UM emphasizes the mobility and dispersion of manufacturing resources and users [25]. In addition, the original definition of UM implied that a system can supply products ubiquitously, which does not necessarily rely on the prevalence of the Internet [26]. The focus is on logistics. However, with advances in ubiquitous technologies such as 3D printing, radio frequency identification (RFID), global positioning system (GPS), autonomous industrial mobile robots, wearable devices, and wireless sensor networks, it is now possible to manufacture products ubiquitously, leading to the establishment of a UM system. Some statistics on the applications of these ubiquitous technologies to UM is provided in Fig. 1.3. According to this figure, in the last 20 years, the number of related references discussing the applications of 3D printing to UM is the largest.

In addition, converting recipes or manufacturing execution plans helps transfer the production of orders and is considered an essential step for UM [27]. Furthermore, distant operation and virtual control of equipment such as robotic arms, computer numerical control (CNC) machines, and 3D printers have provided several opportunities for UM. In particular, 3D printers are cheap to acquire and easy to set up. The prevalence of 3D printing facilities and resources is conducive to the formation of a UM network. The combination of 3D printing (i.e., additive manufacturing) and CNC machining (i.e., subtractive manufacturing) is called hybrid manufacturing [28]. A UM system comprising both additive and subtractive manufacturing equipment is therefore a ubiquitous hybrid manufacturing system.

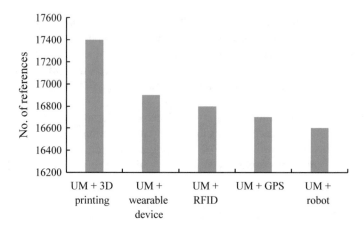

Fig. 1.3 The number of references about ubiquitous technology applications to UM from 2000 to 2020 (data source: Google Scholar)

1.3   Various Types of Ubiquitous Manufacturing Systems

There are various types of UM systems. The original type is a logistics system that supplies a product everywhere [26]. One of the most common types is the application of ubiquitous computing technologies to a manufacturing system [29]. For example, Huang et al. and Zhang et al. defined a UM system as a wireless sensor network that uses RFID tags and receivers, automatic identification (auto-ID) sensors, and wireless information or communication networks to automatically collect, synchronize, and/or process manufacturing data [30, 31]. According to this definition, UM is the same as wireless manufacturing or e-manufacturing. However, this type of UM system is usually confined to the operations of a single factory or logistics system.

Another type of UM system deploys manufacturing resources, services, and/or facilities that use the same raw materials, and produces comparable products as ubiquitously as possible, which is called a ubiquitous industry [26]. According to the threshold set by Alexandersson [32], in a ubiquitous industry, facilities can be found in each region with a population of at least 10,000. In the United States, only two industries, printing and publishing and food processing, met this threshold. The focus is on internationalization and distributivity. To achieve these, a hypersized manufacturing network is usually required. A concern with such a manufacturing network is the balance between the locational costs of production and the distribution costs of finished goods. An alternative is to form a UM networked system through cross-factory or cross-enterprise collaboration. Putnik et al. [33] defined a UM system as a collection of intelligent devices, logically and/or spatially distributed, that change dynamically and reconfigure automatically for new tasks through the provision of manufacturing services, and are supported by semantic tools for unambiguous communication. Such a UM system is a cyberphysical system (CPS) that utilizes Web (or cloud) services, so as to elevate the scalability of the UM system [27]. In the view of Wang et al. [24], a UM system is an elastic and economic service-oriented production model.

Wang et al. [24] classified existing cloud manufacturing applications into two categories: applications of cloud computing technologies to manufacturers and cloud-based manufacturing systems. Existing UM applications can be classified in the same manner. However, a UM system may not rely on the application of cloud or ubiquitous computing technologies. A comparison of various types (or definitions) of UM systems is provided in Table 1.1. UM systems based on the application of 3D printing are the focus of this book, while those based on the deployment of RFIDs are not. Such UM systems belong to the first, second, and fifth types in Table 1.1.

Table 1.1 Various types of UM systems

Reference	Definition	Focus
Foust [26]	A UM system is a logistics system that supplies a product everywhere	Logistics
Foust [26]	A UM system deploys manufacturing resources, services, and/or facilities that use the same raw materials, and produces comparable products as ubiquitously as possible	Internationalization and distributivity
Wang et al. [24]	A UM system is an elastic and flexible service-oriented production model	Information service
Huang et al. [29]; Zhang et al. [31]	A UM system is a wireless sensor network that uses RFID tags and receivers, auto-ID sensors, and wireless information or communication networks to automatically collect, synchronize, and/or process manufacturing data	Ubiquitous technology applications
Putnik et al. [33]	A UM system is a collection of intelligent devices, logically and/or spatially distributed, that change dynamically and reconfigure automatically for new tasks through the provision of manufacturing services, and are supported by semantic tools for unambiguous communication	Cross-factory or cross-enterprise collaboration

Table 1.2 Comparison of UM systems based on various ubiquitous technology applications

Ubiquitous technology	Functionality	Costs	Human intervention
RFID	Information gathering	Very low	Not required
Robots	Transportation	Very high	Not required
CNC machines	Prototyping, manufacturing	Very high	Not required
3D Printers	Prototyping, manufacturing	Low–very high	Required

Most types of UM systems comprise highly automatic and information technology (IT)-intensive facilities that can quickly respond to production conditional or environmental changes. However, a conventional manufacturer can also benefit through utilizing the service of a UM system. In addition, many problems of a factory cannot be solved simply by resorting to more ITs or advanced automation. For example, a scenario in cloud manufacturing (or UM) is to monitor the real-time performance on a shop floor with range cameras, sensors, smart meters, and device controllers [24]. However, the captured images will be very difficult to analyze if the shop floor is messy or operators do not standardize their operations. In such situations, traditional system improvement activities, such as 5s, are more effective than the acquisition of new IT systems or services. For these reasons, dwelling into the details of IT is not necessarily a promising way to solve the problems of a conventional manufacturer. Nevertheless, the benefits of cloud manufacturing (or UM) include cost savings, efficiency, additional data analysis capabilities, advanced automation (such as cloud robotics), flexibility, and closer partner relationships [24, 27]. A comparison of UM systems based on various ubiquitous technology applications is provided in Table 1.2.

1.4 Application of Three-Dimensional Printing to Ubiquitous Manufacturing

Although UM is an effective means of elevating the scalability of a manufacturing system, interoperability is still difficult to achieve [34]. 3D printing provides a solution to this problem. Specifically, the ease and low cost of using a 3D printer and the convenience of exchanging 3D models online are cultivating 3D printing-based UM systems [2, 6, 35]. By contrast, most conventional machines, unlike 3D printers, are dedicated to specific products and may not be suitable for supporting UM [6]. For example, websites like My Mini Factory (www.myminifactory.com) and shapeways (www.shapeways.com) gather 3D models from volunteers worldwide and make them accessible from any place and at any time. If a customer does not have a 3D printer, websites such as 3D Hubs (www.3dhubs.com) help to find a 3D printing facility close to the customer to print a 3D model. A UM system based on 3D printing is actually an application of e-commerce or mobile commerce to 3D printing.

3D printing has been applied to aircraft manufacturing, maintenance, repair, and overhaul for decades [36]. For example, more than 1000 parts in an Airbus A350 have been 3D printed with ULTEM 9085 resin using fused deposition modeling technology [37]. Large, multiple-printer 3D printing systems have been built to realize rapid manufacturing [38]. An efficient UM system is also required to expedite the printing process [37].

Lin and Chen [39] constructed a UM networked system, in which a customer places an order for an action figure by using a client-side app or a Web-based interface and pays online. The system server then assigns the order to the 3D printing facility nearest the customer to print the required action figure. When a 3D printing facility receives an instruction to print the action figure, it automatically generates homogeneous slice contours, selects the feature contours in a representative layer, computes the optimal build direction, and assigns control functions to keep a balance among model accuracy, material consumption, and the printing time [40]. Subsequently, the customer is informed of the location and route to the recommended 3D printing facility to pick up the printed action figure.

Chen [41] constructed another UM system for a similar purpose, in which an order is composed of multiple pieces of an action figure. An efficient production plan is, therefore, to distribute the required pieces among multiple 3D printing facilities to fulfill the order collaboratively. After printing, a freight vehicle visits each 3D printing facility to collect the printed pieces and deliver them to the customer.

3D printing has great potential for establishing a ubiquitous service in the dental industry [42]. Fabricating a denture in the traditional way takes about 2 to 3 weeks, mostly owing to the shortage of available denturists. The application of 3D printing to fabricating dental parts dates back to the early 2000s. In the beginning, 3D printing was applied to the production of a single dental implant, which was done at a single 3D printing facility and did not need the collaboration among multiple 3D printing facilities [43]. However, when a 3D printing facility is busy, all unprocessed jobs have to wait, which results in delays in delivering orders to customers (i.e., doctors or patients). A UM system composed of multiple 3D printing facilities is able to solve this problem. For this purpose, Wang et al. [42] established a UM system for making dental parts using 3D printing. Their UM system is a composite client-server system that distributes the required pieces of an order among several 3D printing facilities to fulfill the order collaboratively. Different from static service networks such as supply chains, only 3D printing facilities close to a customer are considered. As a result, participating 3D printing facilities vary from customer to customer and are unknown in advance, resembling the characteristics of a cloud-based service system.

A UM system based on 3D printers is different from those based on robots or CNC machines. After receiving a production task from the UM system administrator, a 3D printer still needs to be set up manually to optimize the printing performance, while a robot or CNC machine can be directly controlled by the instruction codes associated with the production task [24].

1.5 Organization of This Book

This book aims to introduce the application of 3D printing to UM, including capacity and production planning models, decision-making methodologies, system architectures, and applications. In specific, the outline of the present book is structured as follows.

In the current chapter, first, the definitions of 3D printing and UM are given. Then, existing UM systems are classified. Some applications of 3D printing to UM are also reviewed.

Chapter 2, Application of Ubiquitous Manufacturing to a Conventional Manufacturer, describes how a conventional manufacturer that is not highly automated can benefit from the application of UM. Under a UM environment, a conventional manufacturer can choose among various types of capacity in production planning. A conventional manufacturer usually utilizes self-owned capacity and foundry capacity first to minimize total costs, and resorts to cloud-based capacity if actual demand cannot be fully met by utilizing self-owned and foundry capacity.

Chapter 3, Three-dimensional Printing Capacity Planning, describes how a manufacturer (or 3D printing facility) can build up its 3D printing capacity. Although a 3D printer is most suitable for fabricating specific types of 3D objects (or parts), it theoretically can print almost all types of 3D objects, making the selection of an optimal 3D printer a difficult task. If multiple 3D printers are to be acquired, a capacity planner can minimize the risk by selecting a set of printers with maximal diversity, so as to respond to unexpected or changing demand. In this chapter, first, the application of an analytic hierarchy process (AHP) approach to choose the most suitable 3D printer for a manufacturer is introduced. Subsequently, a decomposition AHP approach is introduced and applied to choose multiple 3D printers that are as diverse as possible by discovering the multiple viewpoints held by a capacity planner.

Chapter 4, Capacity Planning for a Ubiquitous Manufacturing System Based on Three-dimensional Printing, describes how to choose suitable 3D printing facilities for establishing a UM system by considering available capacity, transportation time, costs, product quality, and partner relationships. To this end, an AHP and the technique for order preference by similarity to the ideal solution (TOPSIS) approach is introduced.

In Chapter 5, Production and Transportation Planning for a Ubiquitous Manufacturing System Based on Three-dimensional Printing, decision-making models are formulated and optimized for planning the production and transportation of a UM system based on 3D printing, so as to balance the workloads on 3D printing facilities and identify the shortest delivery path.

Chapter 6, Quality Control in a 3D printing-based UM System, discusses how to evaluate the quality of a 3D-printed item, and how to enhance product quality through taking quality control (QC) actions in a UM system based on 3D printing. In this chapter, the current practices in a 3D printing-based UM system are mapped to the stages of a QC cycle (i.e., product design, process planning, incoming quality control, in-process quality control, and outgoing quality control) to find out shortages of the current practices and opportunities that can be explored in the future.

References

1. H. Lipson, M. Kurman, *Fabricated: the new world of 3D printing* (Wiley, NJ, 2013)
2. T. Chen, Y.C. Lin, Feasibility evaluation and optimization of a smart manufacturing system based on 3D printing. Int. J. Intell. Syst. **32**, 394–413 (2017)
3. H. Kodama, A scheme for three-dimensional display by automatic fabrication of three-dimensional model. IEICE Trans. Electron. (Japanese Edition) J64-C(4), 237–41 (1981)
4. H. Kodama, A scheme for three-dimensional display by automatic fabrication of three-dimensional model. IEICE Trans. Electron. **J64-C**(4), 237–241 (2013)
5. H.N. Chia, B.M. Wu, Recent advances in 3D printing of biomaterials. J. Biol. Eng. **9**(1), 4 (2015)
6. T. Chen, Y.C. Wang, An advanced IoT system for assisting ubiquitous manufacturing with 3D printing. Int. J. Adv. Manuf. Technol. **103**(5–8), 1721–1733 (2019)
7. R. Harris, The 7 Categories of Additive Manufacturing (2012). https://www.lboro.ac.uk/research/amrg/about/the7categoriesofadditivemanufacturing/
8. All3DP, 2020 Types of 3D Printing Technology (2020). https://all3dp.com/1/types-of-3d-printers-3d-printing-technology/
9. A. Kochan, Magnetic pulse welding shows potential for automotive applications. Assem. Autom. **20**, 129–132 (2000)
10. R. Bogue, 3D printing: the dawn of a new era in manufacturing? Assembly Autom. **33**, 307–311 (2013)
11. C.L. Ventola, Medical applications for 3D printing: current and projected uses. Pharm. Ther. **39**, 704–711 (2014)
12. F. Zhang, S. Liu, P. Zhao, T. Liu, J. Sun, Titanium/nanodiamond nanocomposites: effect of nanodiamond on microstructure and mechanical properties of titanium. Mater. Des. **131**, 144–155 (2017)
13. F. Rengier, A. Mehndiratta, H. von Tengg-Kobligk, C.M. Zechmann, R. Unterhinninghofen, H.-U. Kauczor, F.L. Giesel, 3D printing based on imaging data: review of medical applications. Int. J. Comput. Assist. Radiol. Surg. **5**, 335–341 (2010)
14. T. Grünberger, R. Domröse, Direct metal laser sintering. Laser Tech. J. **12**(1), 45–48 (2015)
15. Y.-C. Wang, T. Chen, Y.-L. Yeh, Advanced 3D printing technologies for the aircraft industry: a fuzzy systematic approach for assessing the critical factors. Int. J. Adv. Manuf. Technol. **105**, 4059–4069 (2019)
16. M.J. Donachie Jr, *Titanium. A Technical Guide* (ASM International, Metals Park, OH, 1988)
17. B. Smith, The boeing 777. Adv. Mater. Process. **161**(9), 41–44 (2003)
18. G. Cesaretti, E. Dini, X. De Kestelier, V. Colla, L. Pambaguian, Building components for an outpost on the Lunar soil by means of a novel 3D printing technology. Acta Astronaut. **93**, 430–450 (2014)
19. The Economist, A Third Industrial Revolution (2012), http://www.economist.com/node/21552901
20. H. Kim, Y. Lin, T.L.B. Tseng, A review on quality control in additive manufacturing. Rapid Prototyp. J. **24**(3), 645–669 (2018)
21. H.C. Wu, T.C.T. Chen, Quality control issues in 3D-printing manufacturing: a review. Rapid Prototyp. J. **24**(3), 607–614 (2018)
22. S. Klein, M. Avery, G. Adams, S. Pollard, S. Simske, From scan to print: 3D printing as a means for replication, in *NIP & Digital Fabrication Conference* (2014), No. 1, pp. 417–421
23. T. Chen, H.R. Tsai, Ubiquitous manufacturing: current practices, challenges, and opportunities. Robot. Comput.-Integr. Manuf. **45**, 126–132 (2017)

24. X.V. Wang, L. Wang, A. Mohammed, M. Givehchi, Ubiquitous manufacturing system based on cloud: a robotics application. Robot. Comput.-Integr. Manuf. **45**, 116–125 (2017)
25. T.C.T. Chen, T.W. Liao, D.H. Lee, G. Bocewicz, Ubiquitous manufacturing. Robot. Comput.-Integr. Manuf. **45**, 1–2 (2017)
26. B.J. Foust, Ubiquitous manufacturing. Ann. Assoc. Am. Geogr. **65**(1), 13–17 (1975)
27. T. Chen, Strengthening the competitiveness and sustainability of a semiconductor manufacturer with cloud manufacturing. Sustainability **6**, 251–268 (2014)
28. S. Hendrixson, AM 101: Hybrid Manufacturing (2019), https://www.additivemanufacturing.media/blog/post/am-101-hybrid-manufacturing
29. G. Putnik, Advanced manufacturing systems and enterprises: cloud and ubiquitous manufacturing and an architecture. J. Appl. Eng. Sci. **10**(3), 127–143 (2012)
30. G.Q. Huang, P.K. Wright, S.T. Newman, Wireless manufacturing: a literature review, recent developments, and case studies. Int. J. Comput. Integr. Manuf. **22**(7), 579–594 (2009)
31. Y.F. Zhang, T. Qu, Q. Ho, G.Q. Huang, Real-time work-in-progress management for smart object enabled ubiquitous shop floor environment. Int. J. Comput. Integr. Manuf. **24**(5), 431–445 (2011)
32. S.K. Cha, J.Y. Song, J.S. Choi, u-Manufacturing needs M2M (machine to machine) for ubiquitous computing world, in *International Conference on Control, Automation and Systems* (2005), pp. 1–3
33. G. Putnik, C. Cardeira, P. Leitão, F. Restivo, J. Santos, A. Sluga, P. Butala, Towards ubiquitous production systems and enterprises, in *IEEE International Symposium on Industrial Electronics* (2007), pp. 3203–3208
34. G. Alexandersson, *The Industrial Structure of American Cities* (The University of Nebraska Press, Lincoln, 1965)
35. L. Zimmermann, T. Chen, K. Shea, A 3D, performance-driven generative design framework: automating the link from a 3D spatial grammar interpreter to structural finite element analysis and stochastic optimization. Artif. Intell. Eng. Des. Anal. Manuf. **32**, 189–199 (2018)
36. C.-W. Lin, T. Chen, 3D printing technologies for enhancing the sustainability of an aircraft manufacturing or MRO company—a multi-expert partial-consensus FAHP analysis. Int. J. Adv. Manuf. Technol. **105**, 4171–4180 (2019)
37. S. Helsel, Airbus' New A350 XWB Aircraft Contains Over 1,000 3D-printed Parts (2017), http://inside3dprinting.com/news/airbus-new-a350-xwb-aircraft-contains-over-1000-3d-printed-parts/30291/
38. L.E. Murr, Frontiers of 3D printing/additive manufacturing: from human organs to aircraft fabrication. J. Mater. Sci. Technol. **32**(10), 987–995 (2016)
39. Y.-C. Lin, T. Chen, A ubiquitous manufacturing network system. Robot. Comput.-Integr. Manuf. **45**, 157–167 (2017)
40. Z. Zhang, X. Wang, X. Zhu, Q. Cao, F. Tao, Cloud manufacturing paradigm with ubiquitous robotic system for product customization. Robot. Comput.-Integr. Manuf. **60**, 12–22 (2019)
41. T. Chen, Fuzzy approach for production planning by using a three-dimensional printing-based ubiquitous manufacturing system. Artif. Intell. Eng. Des. Anal. Manuf. **33**(4), 458–468 (2019)
42. Y.-C. Wang, T. Chen, Y.-C. Lin, A collaborative and ubiquitous system for fabricating dental parts using 3D printing technologies. Healthcare **7**, 103 (2019)
43. I.J. Petrick, T.W. Simpson, 3D printing disrupts manufacturing: how economies of one create new rules of competition. Res. Technol. Manag. **56**, 12–16 (2013)

Chapter 2
Application of Ubiquitous Manufacturing to a Conventional Manufacturer

2.1 Migrating to Three-Dimensional Printing

In the past, a conventional manufacturer implemented ubiquitous manufacturing (UM) in the following ways:

(1) Cooperating with a logistics network to supply products ubiquitously.
(2) Building factories in as many places as possible, especially those close to customers.
(3) Collaborating with foundries in as many places as possible, especially those close to customers.
(4) Use a mix of the previous methods.

New opportunities emerge with the prevalent applications of ubiquitous computing technologies. For example, factories can share their unused capacity with each other online with the aid of a UM service provider. In this way, factories short of capacity can deliver more orders to make more profits, while lowly utilized factories can exchange their capacity for revenues. In this scenario, cooperative relationships are more temporary and informal, which is different from the past.

This chapter is dedicated to the mid-term or long-term production planning for a conventional manufacturer under a UM environment, which has rarely been investigated in the past studies. In contrast, most existing methods are devoted to the short-term production planning or job scheduling in a UM system [1], which assumes the extensive adoption of radio frequency identification (RFID) sensors, a cross-organizational information technology (IT) system architecture, and ubiquitous computing methodologies. Such an assumption may be impractical to some conventional manufacturers. Nevertheless, a conventional manufacturer can still benefit from migrating to 3D printing, resorting to cloud-based capacity, or selectively deploying RFID sensors, which is much easier and practical.

A conventional manufacturer can consider migrating its manufacturing capacity to three-dimensional (3D) printing by addressing the following concerns:

© The Author(s), under exclusive license to Springer Nature Switzerland AG 2020
T.-C. Chen, *3D Printing and Ubiquitous Manufacturing*,
SpringerBriefs in Applied Sciences and Technology,
https://doi.org/10.1007/978-3-030-49150-5_2

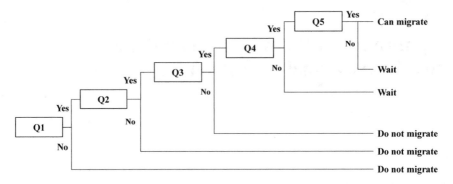

Fig. 2.1 The decision tree for migrating to 3D printing

- Are the 3D models of products available?
- Can the 3D models be converted into the STL file format?
- Are the materials for producing products acceptable to a 3D printer?
- Can the required precision for products be achieved by a 3D printer?
- Is the budget on 3D printers to meet the daily throughput acceptable?

    A decision tree is provided in Fig. 2.1 to assist make this decision-making.

2.2 Applicability of Ubiquitous Manufacturing to a Conventional Manufacturer

The procedure for planning the mid-term or long-term production of a conventional manufacturer under a UM environment comprises the following steps:

Step 1. Determine the planning horizon.
Step 2. Forecast the demand for a product within each future period.
Step 3. Estimate the unit cost and yield of the product within every future period.
Step 4. Calculate the number of machines required if all the demand is fulfilled by itself.
Step 5. Collect the information about the availability and costs of foundry and cloud-based capacity within each future period.
Step 6. Formulate and optimize the production planning model to switch between self-owned capacity and foundry capacity to minimize total costs.
Step 7. Resort to cloud-based capacity if actual demand cannot be fully met by utilizing self-owned and foundry capacity.

    The flowchart in Fig. 2.2 illustrates this procedure.

Fig. 2.2 The procedure for planning the mid-term or long-term production of a conventional manufacturer under a UM environment

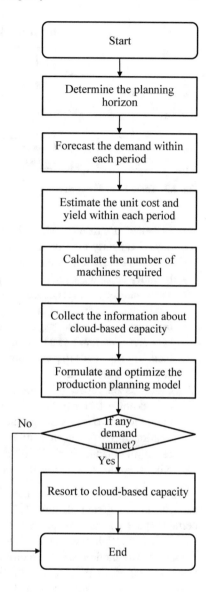

2.3 Production Planning by Prioritizing Self-Owned Capacity

Theoretically, there are three types of capacity that can be utilized by a manufacturer under a UM environment, as illustrated in Fig. 2.3:

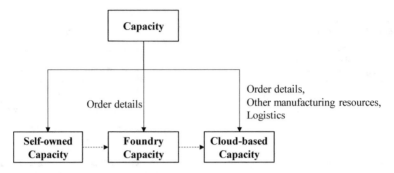

Fig. 2.3 Three types of capacity under a UM environment

(1) Self-owned capacity, which is the capacity owned by a manufacturer. Self-owned capacity may be located in a single factory or distributed over several factories controlled by the manufacturer [2].

(2) Foundry capacity, which is the capacity of foundries that can be utilized within a certain time interval according to the formal agreement between foundries and a manufacturer [3]. Two major types of foundries are original equipment manufacturers (OEMs) and original design manufacturers (ODMs). The relationship between a foundry and a manufacturer is usually long-term. As a result, a manufacturer only needs to pass order details to foundries. The required coordination is largely simplified.

(3) Cloud-based capacity, which is the available capacity information provided by unknown factories (i.e., the capacity providers) and can be accessed by a manufacturer online through the intervention of a cloud service provider. The relationship between capacity providers and a manufacturer is temporary. As a result, the manufacturer is still responsible for the preparation and logistics of other manufacturing resources [4].

Under a UM environment, a manufacturer has to choose among various types of capacity in production planning [5]. Since self-owned capacity is usually the most certain capacity, a manufacturer will first utilize self-owned capacity to meet the forecasted demand. Self-owned capacity-based manufacturing is also the most efficient manufacturing mode because no cross-organizational collaboration and transportation are required. To this end, first the demand for a product within each period needs to be forecasted. A number of methods have been proposed to fulfill this purpose, e.g., nonlinear regression, artificial neural networks, etc. [6, 7]. However, demand is subject to much uncertainty. To tackle the uncertainty, the demand within a period can be forecasted with a triangular fuzzy number (TFN) [8]:

$$\tilde{d}(t) = (d_1(t), d_2(t), d_3(t)) \tag{2.1}$$

as shown in Fig. 2.4. The membership function of $\tilde{d}(t)$ is

Fig. 2.4 Forecasting the
demand for a product with a
TFN

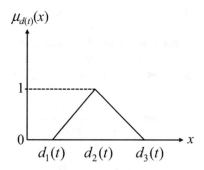

$$\mu_{\tilde{d}(t)}(x) = \begin{cases} \frac{x - d_1(t)}{d_2(t) - d_1(t)} & \text{if} & d_1(t) \le x < d_2(t) \\ \frac{x - d_3(t)}{d_2(t) - d_3(t)} & \text{if} & d_2(t) \le x < d_3(t) \\ 0 & \text{otherwise} \end{cases} \tag{2.2}$$

The formulae for some arithmetic operations on TFNs are provided as follows:

- Fuzzy addition:

$$\widetilde{A}(+)\widetilde{B} = (A_1 + B_1, A_2 + B_2, A_3 + B_3) \tag{2.3}$$

- Fuzzy subtraction:

$$\widetilde{A}(-)\widetilde{B} = (A_1 - B_3, A_2 - B_2, A_3 - B_1) \tag{2.4}$$

- Fuzzy scalar multiplication:

$$k(\cdot)\widetilde{A} = (kA_1, kA_2, kA_3) \text{ if } k \ge 0 \tag{2.5}$$

- Fuzzy multiplication:

$$\widetilde{A}(\times)\widetilde{B} = (A_1 B_1, A_2 B_2, A_3 B_3) \text{ if } A_1, B_1 \ge 0 \tag{2.6}$$

- Fuzzy division:

$$\widetilde{A}(/)\widetilde{B} = (A_1/B_3, A_2/B_2, A_3/B_1) \text{ if } A_1 \ge 0, B_1 > 0 \tag{2.7}$$

Subsequently, the yield and unit cost of the product when demand is fully self-made can be estimated. The product yield within period t is forecasted as $\tilde{y}(t)$; $\tilde{y}(t) \in [0, 1]$. The unit cost within period t is forecasted as $\tilde{c}_s(t)$. Both are given in TFNs. Fuzzy methods for forecasting the yield of a product include fuzzy linear regression, fuzzy grading, adaptive neuro-fuzzy inference system (ANFIS) networks, and others [9–11]. For forecasting the unit cost of a product, applicable methods include fuzzy multi-attribute utility theory, fuzzy linear regression, fuzzy inference systems, etc. [12–14].

Since self-owned capacity is usually the most certain capacity, a manufacturer will build up and utilize self-owned capacity first. The required number of machines that meet the forecasted demand within each period can be determined as

$$
\begin{aligned}
\tilde{m} &= \max_t \left\lceil \frac{p(\cdot)\tilde{d}(t)}{\tilde{y}(t)(\times)\tilde{v}(t)(\cdot)W(t)} \right\rceil \\
&= \max_t \left\lceil \left(\frac{pd_1(t)}{y_3(t)v_3(t)W(t)}, \frac{pd_2(t)}{y_2(t)v_2(t)W(t)}, \frac{pd_3(t)}{y_1(t)v_1(t)W(t)} \right) \right\rceil \\
&= \left(\max_t \left\lceil \frac{pd_1(t)}{y_3(t)v_3(t)W(t)} \right\rceil, \max_t \left\lceil \frac{pd_2(t)}{y_2(t)v_2(t)W(t)} \right\rceil, \max_t \left\lceil \frac{pd_3(t)}{y_1(t)v_1(t)W(t)} \right\rceil \right)
\end{aligned}
$$
(2.8)

According to the arithmetic for TFNs, p is the unit processing time. $\tilde{v}(t)$ indicates the forecasted availability within period t; $\tilde{v}(t) \in [0,\ 1]$. $W(t)$ is the working hours within period t.

Example 2.1

A metal cutting machine is used to manufacture a product in a factory. The unit processing time (p) is 0.73 h/product; the monthly usage cost of the machine (U) is 2200 USD; the unit variable cost of the product (c_1) is 25 USD per piece. The planning horizon (T) is 12 months. The demand for the product, the working hours within each period, the yield of the product, and the availability of the machine within each period have been collected or forecasted, as summarized in Table 2.1.

Table 2.1 The collected or forecasted data

t	$\tilde{d}(t)$ (pieces)	$W(t)$ (hrs)	$\tilde{y}(t)$	$\tilde{v}(t)$
1	(970, 994, 1030)	744	(71%, 75%, 77%)	(73%, 75%, 82%)
2	(1380, 1499, 1635)	672	(71%, 76%, 78%)	(76%, 80%, 86%)
3	(1175, 1266, 1362)	744	(72%, 77%, 79%)	(83%, 85%, 90%)
4	(1656, 1729, 1818)	720	(72%, 78%, 81%)	(83%, 90%, 92%)
5	(2016, 2117, 2247)	744	(73%, 78%, 82%)	(82%, 90%, 98%)
6	(2350, 2498, 2650)	720	(74%, 79%, 82%)	(89%, 90%, 91%)
7	(2182, 2313, 2452)	744	(75%, 79%, 83%)	(90%, 90%, 92%)
8	(1767, 1825, 1900)	744	(75%, 79%, 83%)	(87%, 90%, 93%)
9	(1697, 1755, 1837)	720	(76%, 80%, 82%)	(84%, 90%, 91%)
10	(1457, 1527, 1627)	744	(77%, 80%, 86%)	(83%, 90%, 95%)
11	(2366, 2495, 2678)	720	(78%, 80%, 86%)	(87%, 90%, 96%)
12	(2085, 2192, 2343)	744	(79%, 81%, 86%)	(88%, 90%, 96%)

$$\tilde{u}(t) = \dfrac{\dfrac{p(\cdot)\tilde{d}(t)}{\bar{y}(t)(\times)\bar{v}(t)(\times)W(t)}}{m}$$

$$= \left(\frac{pd_1(t)}{my_3(t)v_3(t)W(t)}, \frac{pd_2(t)}{my_2(t)v_2(t)W(t)}, \frac{pd_3(t)}{my_1(t)v_1(t)W(t)} \right) \qquad (2.12)$$

In Eq. (2.12), $p(\cdot)\tilde{d}(t)/\bar{y}(t)(\times)\bar{v}(t)(\times)W(t)$ is the number of machines required, while m is the number of machines acquired. Dividing the former by the latter gives the expected utilization.

Example 2.2
In the previous example, if only three units of the machine are acquired, the forecasted quantity of self-made products within the first period can be calculated as

$$\tilde{x}_s(1) = \left(\min\left(970, \left\lfloor \frac{3 \cdot 71\% \cdot 73\% \cdot 744}{0.73} \right\rfloor \right), \min\left(994, \left\lfloor \frac{3 \cdot 75\% \cdot 75\% \cdot 744}{0.73} \right\rfloor \right), \right.$$
$$\left. \min\left(1030, \left\lfloor \frac{3 \cdot 77\% \cdot 82\% \cdot 744}{0.73} \right\rfloor \right) \right)$$
$$= (970, 994, 994)$$

The quantity of products that needs to be made by utilizing foundry capacity is

$$\tilde{x}_f(1) = (\max(970 - 994, 0), \max(994 - 994, 0), \max(1030 - 970, 0))$$
$$= (0, 0, 60)$$

The results within the other periods are summarized in Table 2.2.

Table 2.2 The forecasted quantities of products made in various ways	t	$\tilde{x}_s(t)$ (pieces)	$\tilde{x}_f(t)$ (pieces)
	1	(970, 994, 994)	(0, 0, 60)
	2	(1380, 1499, 1499)	(0, 0, 255)
	3	(1175, 1266, 1266)	(0, 0, 187)
	4	(1656, 1729, 1729)	(0, 0, 162)
	5	(1839, 2117, 2117)	(0, 0, 408)
	6	(1957, 2103, 2216)	(134, 395, 693)
	7	(2060, 2173, 2313)	(0, 140, 392)
	8	(1767, 1825, 1825)	(0, 0, 133)
	9	(1697, 1755, 1755)	(0, 0, 140)
	10	(1457, 1527, 1527)	(0, 0, 170)
	11	(2020, 2130, 2453)	(0, 365, 658)
	12	(2085, 2192, 2192)	(0, 0, 258)

The number of machines required by the factory is determined according Eq. (2.8) as

$$
\begin{aligned}
\tilde{m} = \Bigg(& \max_t \left\lceil \frac{0.73 \cdot 970}{77\% \cdot 82\% \cdot 744}, \ \cdots, \ \frac{0.73 \cdot 2085}{86\% \cdot 96\% \cdot 744} \right\rceil, \\
& \max_t \left\lceil \frac{0.73 \cdot 994}{75\% \cdot 75\% \cdot 744}, \ \cdots, \ \frac{0.73 \cdot 2192}{81\% \cdot 90\% \cdot 744} \right\rceil, \\
& \max_t \left\lceil \frac{0.73 \cdot 1030}{71\% \cdot 73\% \cdot 744}, \ \cdots, \ \frac{0.73 \cdot 2343}{79\% \cdot 88\% \cdot 744} \right\rceil \Bigg) \\
= & (4, 4, 5)
\end{aligned}
$$

Therefore, if the manufacturer acquires four to five (or more) units of the machine all the demand can be self-made. Otherwise, part of the demand needs to be made by utilizing foundry capacity.

In Eq. (2.8), $p(\cdot)\tilde{d}(t)$ is the required self-owned capacity within period t, while $\tilde{y}(t)(\times)\tilde{v}(t)(\cdot)W(t)$ is the capacity provided by a single machine. Dividing the former by the latter gives the required number of machines. If m machines are actually acquired, the number of self-made products within period t, $\tilde{x}_s(t)$, is forecasted as

$$
\begin{aligned}
\tilde{x}_s(t) = & \min\left(\tilde{d}(t), \ \left\lfloor \frac{m(\cdot)\tilde{y}(t)(\times)\tilde{v}(t)(\cdot)W(t)}{p} \right\rfloor \right) \\
= & \left(\min\left(d_1(t), \ \left\lfloor \frac{my_1(t)v_1(t)W(t)}{p} \right\rfloor \right), \ \min\left(d_2(t), \ \left\lfloor \frac{my_2(t)v_2(t)W(t)}{p} \right\rfloor \right), \right. \\
& \left. \min\left(d_3(t), \ \left\lfloor \frac{my_3(t)v_3(t)W(t)}{p} \right\rfloor \right) \right)
\end{aligned}
\tag{2.9}
$$

In Eq. (2.9), $\lfloor m(\cdot)\tilde{y}(t)(\times)\tilde{v}(t)(\cdot)W(t)/p \rfloor$ is the quantity of products that can be made on m machines within period t, which should not be more than $\tilde{d}(t)$. If $m < \tilde{m}$, then the unmet demand within period t is forecasted as

$$
\begin{aligned}
\tilde{\delta}(t) = & \max(\tilde{d}(t)(-)\tilde{x}_s(t), 0) \\
= & (\max(d_1(t) - x_{s3}(t), 0), \max(d_2(t) - x_{s2}(t), 0), \max(d_3(t) - x_{s1}(t), 0))
\end{aligned}
\tag{2.10}
$$

which is to be handled by utilizing foundry capacity:

$$
\tilde{x}_f(t) = \tilde{\delta}(t)
\tag{2.11}
$$

$\tilde{x}_f(t)$ is the quantity of products made by utilizing foundry capacity. As a result, the required foundry capacity within period t is $p\tilde{x}_f(t)$. In addition, the expected utilization can be derived as

2.4 Production Planning by Considering Both Self-Owned and Foundry Capacity

In the previous section, the priority of self-owned capacity is absolutely higher than that of foundry capacity. A manufacturer may seek a better balance between self-owned capacity and foundry capacity to optimize the cost-effectiveness by minimizing the sum of total costs:

$$\text{Min } \tilde{Z}_1 = \sum_{t=t}^{T} \tilde{C}(t) \tag{2.13}$$

subject to

$$\tilde{x}_s(t)(+)\tilde{x}_f(t) = \tilde{d}(t) \tag{2.14}$$

where

$$\tilde{x}_s(t) \leq \frac{m(\cdot)\tilde{y}(t)(\times)\tilde{v}(t)(\cdot)W(t)}{p} \tag{2.15}$$

In Eq. (2.14), the quantity of products made by utilizing self-owned capacity plus that made by utilizing foundry capacity is equal to the forecasted demand. Constraint (2.15) requests that the quantity of self-made products be fewer than that can be made. The forecasted unit cost within period t can be calculated as

$$\tilde{c}_s(t) = c_1(+)\frac{mU}{\tilde{x}_s(t)} \tag{2.16}$$

where U is the usage cost of a machine per month. c_1 denotes the unit variable cost. $mU(/)\tilde{x}_s(t)$ is the unit fixed cost that is derived by dividing the equipment usage cost among all self-made products. The forecasted total self-made costs are obtained by multiplying the forecasted unit cost to the quantity of self-made products as

$$\tilde{C}_s(t) = \tilde{c}_s(t)(\times)\tilde{x}_s(t)$$
$$= c_1\tilde{x}_s(t)(+)mU \tag{2.17}$$

while the total foundry costs are forecasted in a similar way:

$$\tilde{C}_f(t) = c_f(\cdot)\tilde{x}_f(t) \tag{2.18}$$

Therefore, the forecasted total costs are the sum of self-made costs and foundry costs:

$$\tilde{C}(t) = \tilde{C}_s(t)(+)\tilde{C}_f(t)$$
$$= c_1\tilde{x}_s(t)(+)mU(+)c_f\tilde{x}_f(t) \tag{2.19}$$

As a result, the following capacity and production planning model is established, which is a fuzzy mixed-integer nonlinear programming (FMILP) problem:

(Capacity and Production Planning Model)

$$\text{Min } \tilde{Z}_1 = \sum_{t=t}^{T} \tilde{C}(t) \tag{2.20}$$

subject to

$$\tilde{C}(t) = c_1 \tilde{x}_s(t)(+)mU(+)c_f \tilde{x}_f(t); t = 1 \sim T \tag{2.21}$$

$$\tilde{x}_s(t)(+)\tilde{x}_f(t) = \tilde{d}(t); t = 1 \sim T \tag{2.22}$$

$$\tilde{x}_s(t) \leq \frac{m(\cdot)\tilde{y}(t)(\times)\tilde{v}(t)(\cdot)W(t)}{p}; t = 1 \sim T \tag{2.23}$$

$$\tilde{x}_s(t), \tilde{x}_f(t), m \in Z^+; t = 1 \sim T \tag{2.24}$$

$$\tilde{C}(t) \in R^+; t = 1 \sim T \tag{2.25}$$

The FMILP problem is converted into an equivalent mixed-integer linear programming (MILP) model to facilitate the problem-solving [15, 16]:

(MILP Model)

$$\text{Min } Z_1 = \sum_{t=t}^{T} \frac{C_1(t) + C_2(t) + C_3(t)}{3} \tag{2.26}$$

subject to

$$C_1(t) = c_1 x_{s1}(t) + mU + c_f x_{f1}(t); t = 1 \sim T \tag{2.27}$$

$$C_2(t) = c_1 x_{s2}(t) + mU + c_f x_{f2}(t); t = 1 \sim T \tag{2.28}$$

$$C_3(t) = c_1 x_{s3}(t) + mU + c_f x_{f3}(t); t = 1 \sim T \tag{2.29}$$

$$\frac{x_{s1}(t) + x_{s2}(t) + x_{s3}(t)}{3} + \frac{x_{f1}(t) + x_{f2}(t) + x_{f3}(t)}{3}$$
$$= \frac{d_1(t) + d_2(t) + d_3(t)}{3}; t = 1 \sim T \tag{2.30}$$

$$x_{s1}(t) \leq \frac{my_1(t)v_1(t)W(t)}{p}; t = 1 \sim T \tag{2.31}$$

$$x_{s2}(t) \leq \frac{my_2(t)v_2(t)W(t)}{p}; t = 1 \sim T \tag{2.32}$$

$$x_{s3}(t) \leq \frac{my_3(t)v_3(t)W(t)}{p}; t = 1 \sim T \tag{2.33}$$

$$x_{s1}(t) \leq x_{s2}(t) \leq x_{s3}(t); t = 1 \sim T \tag{2.34}$$

$$x_{f1}(t) \leq x_{f2}(t) \leq x_{f3}(t); t = 1 \sim T \tag{2.35}$$

$$\tilde{x}_s(t), \tilde{x}_f(t), m \in Z^+; t = 1 \sim T \tag{2.36}$$

$$C_1(t), C_2(t), C_3(t) \in R^+; t = 1 \sim T \tag{2.37}$$

Equations (2.27)–(2.29) are the one-to-one mappings between the TFNs on both sides. Constraints (2.34) and (2.35) define the sequence of the three corners of the corresponding TFNs.

Example 2.3

In the previous example, the MILP problem of the case is built and solved using Lingo, as illustrated in Fig. 2.5. The minimal forecasted total costs are 655,295 USD. To achieve that, the manufacturer needs to acquire three units of the machine and resorts to foundries when there is a capacity shortage. The production plan is shown in Table 2.3. It is noted that there may be multiple optimal solutions to the MINP problem. In addition, the ranges of $\tilde{x}_s$ and $\tilde{x}_f$ give the manufacturer a lot of flexibility.

2.5 Resorting to Cloud-Based Capacity

Actual demand is seldom equal to the forecasted demand. As a result,

(1) If actual demand is less than the forecasted demand, i.e., $a(t) < \tilde{d}(t)$, the manufacturer will utilize less self-owned capacity. Specifically speaking, the quantity made by utilizing foundry capacity $(x_f(t))$ remains because of the signed contract, while that made by utilizing self-owned capacity is reduced to

$$x_s(t) = a(t) - x_f(t) \tag{2.38}$$

In addition, no cloud-based capacity will be utilized, i.e., $x_c(t) = 0$.

```
min=Z1;
Z1=dC1+dC2+dC3+dC4+dC5+dC6+dC7+dC8+dC9+dC10+dC11+dC12;
3*dC1=C11+C12+C13;
...
3*dC12=C121+C122+C123;
C11=25*xs11+2200*m+47*xf11;
C12=25*xs12+2200*m+47*xf12;
C13=25*xs13+2200*m+47*xf13;
...
C121=25*xs121+2200*m+47*xf121;
C122=25*xs122+2200*m+47*xf122;
C123=25*xs123+2200*m+47*xf123;
xs11+xs12+xs13+xf11+xf12+xf13=2994;
...
xs121+xs122+xs123+xf121+xf122+xf123=6620;
xs11<=527*m;
xs12<=573*m;
xs13<=637*m;
...
xs121<=711*m;
xs122<=742*m;
xs123<=845*m;
xs11<=xs12;
xs12<=xs13;
...
xs121<=xs122;
xs122<=xs123;
xf11<=xf12;
xf12<=xf13;
...
xf121<=xf122;
xf122<=xf123;
@gin(xs11); @gin(xs12); @gin(xs13); ...; @gin(xs121); @gin(xs122); @gin(xs123);
@gin(xf11); @gin(xf12); @gin(xf13); ...; @gin(xf121); @gin(xf122); @gin(xf123);
@gin(m);
```

Fig. 2.5 The MILP model

Table 2.3 The production plan

t	$\tilde{x}_s(t)$ (pieces)	$\tilde{x}_f(t)$ (pieces)
1	(970, 994, 994)	(0, 0, 60)
2	(1380, 1499, 1499)	(0, 0, 255)
3	(1175, 1266, 1266)	(0, 0, 187)
4	(1656, 1729, 1729)	(0, 0, 162)
5	(1910, 2117, 2117)	(0, 0, 337)
6	(2038, 2190, 2307)	(43, 308, 612)
7	(2148, 2267, 2313)	(0, 46, 304)
8	(1767, 1825, 1825)	(0, 0, 133)
9	(1697, 1755, 1755)	(0, 0, 140)
10	(1457, 1527, 1527)	(0, 0, 170)
11	(2113, 2228, 2495)	(0, 267, 565)
12	(2085, 2192, 2192)	(0, 0, 258)

(2) Otherwise, the manufacturer will utilize unused self-owned capacity or seek for available cloud-based capacity to fill up the shortage:

$$x_s(t) = \min(a(t) - x_f(t), \ \frac{my(t)v(t)W(t)}{p}) \qquad (2.39)$$

where $a(t) - x_f(t)$ is the quantity of products that needs to be made by utilizing self-owned capacity; $my(t)v(t)W(t)/p$ is the quantity that can be made. The quantity that needs to be made by resorting to cloud-based capacity is

$$x_c(t) = a(t) - x_s(t) - x_f(t) \qquad (2.40)$$

The unused self-owned capacity can then be shared online:

$$\xi(t) = \max(0, \ \left\lfloor \frac{my(t)v(t)W(t)}{p} \right\rfloor - x_s(t)) \qquad (2.41)$$

In this way, the manufacturer becomes a cloud capacity supplier.

The prerequisites for these actions include the availability of cloud service providers and the manufacturer's willingness to share capacity and utilize cloud-based capacity.

Example 2.4
In the previous example, the actual demand within each period is known and shown in Table 2.4, based on which the production plan is adjusted. When actual demand was less than the forecasted demand, the manufacturer utilizes less self-owned capacity as a response. Otherwise, the manufacturer utilizes more self-owned capacity (if available) or resorts to cloud-based capacity. The required cloud-based capacity within

Table 2.4 The actual demand within each period

t	$a(t)$ (pieces)
1	1045
2	1536
3	1290
4	1663
5	2550
6	2451
7	2333
8	1752
9	1912
10	1550
11	2657
12	2208

Table 2.5 The adjusted production plan

t	$x_s(t)$ (pieces)	$x_f(t)$ (pieces)	$x_c(t)$ (pieces)
1	1045	0	0
2	1536	0	0
3	1290	0	0
4	1663	0	0
5	2210	0	340
6	1838	613	0
7	2141	192	0
8	1752	0	0
9	1912	0	0
10	1550	0	0
11	2187	470	0
12	2208	0	0

each period is shown in Table 2.5. The adjusted production plan is also presented in this table.

In Table 2.5, the factory uses very little foundry and cloud-based capacity. This is reasonable since foundry capacity and cloud-based capacity are much more expensive than self-owned capacity. Nevertheless, foundry and cloud-based capacity still contribute to the effectiveness of production planning, because actual demand cannot be fully met without foundry and cloud-based capacity, which incurs penalties and results in the loss of future sales.

References

1. L. Wang, S. Guo, X. Li, B. Du, W. Xu, Distributed manufacturing resource selection strategy in cloud manufacturing. Int. J. Adv. Manuf. Technol. **94**(9–12), 3375–3388 (2018)
2. S. Sweat, S. Niu, M.T. Zhang, Z. Zhang, L. Zheng, Multi-factory capacity planning in semiconductor assembly and test manufacturing with multiple-chip products, in *IEEE International Conference on Automation Science and Engineering* (2006), pp 247–252
3. S. Fore, C.T. Mbohwa, Cleaner production for environmental conscious manufacturing in the foundry industry. J. Eng. Design Technol. **8**(3), 314–333 (2010)
4. L. Zhou, L. Zhang, L. Ren, Modelling and simulation of logistics service selection in cloud manufacturing. Procedia CIRP **72**, 916–921 (2018)
5. P. Argoneto, P. Renna, Supporting capacity sharing in the cloud manufacturing environment based on game theory and fuzzy logic. Enterprise Inform. Syst. **10**(2), 193–210 (2016)
6. C.F. Chien, Y.J. Chen, J.T. Peng, Manufacturing intelligence for semiconductor demand forecast based on technology diffusion and product life cycle. Int. J. Prod. Econ. **128**(2), 496–509 (2010)
7. A. Kochak, S. Sharma, Demand forecasting using neural network for supply chain management. Int. J. Mech. Eng. Robot. Res. **4**(1), 96–104 (2015)
8. D. Huang, T. Chen, M.J.J. Wang, A fuzzy set approach for event tree analysis. Fuzzy Sets Syst. **118**(1), 153–165 (2001)
9. T. Chen, M.-J.J. Wang, A fuzzy set approach for yield learning modeling in wafer manufacturing. IEEE Trans. Semicond. Manuf. **12**(2), 252–258 (1999)

10. O. Taylan, Estimating the quality of process yield by fuzzy sets and systems. Expert Syst. Appl. **38**(10), 12599–12607 (2011)
11. S.Z. Abghari, M. Sadi, Application of adaptive neuro-fuzzy inference system for the prediction of the yield distribution of the main products in the steam cracking of atmospheric gasoil. J. Taiwan Inst. Chem. Eng. **44**(3), 365–376 (2013)
12. P.K. Dean Ting, C. Zhang, B. Wang, A. Deshmukh, B. Dubrosky, Product and process cost estimation with fuzzy multi-attribute utility theory. Eng. Econ. **44**(4), 303–331 (1999)
13. T. Chen, Applying the hybrid fuzzy c means-back propagation network approach to forecast the effective cost per die of a semiconductor product. Comput. Ind. Eng. **61**(3), 752–759 (2011)
14. E.M. Shehab, H.S. Abdalla, Manufacturing cost modelling for concurrent product development. Robot. Comput. Integr. Manuf. **17**(4), 341–353 (2001)
15. H.J. Zimmermann, *Fuzzy Set Theory and Its Applications* (Springer, New York, 1991)
16. R. Kumar, S.A. Edalatpanah, S. Jha, R. Singh, A Pythagorean fuzzy approach to the transportation problem. Complex Intell. Syst. **5**(2), 255–263 (2019)

Chapter 3
Three-Dimensional Printing Capacity Planning

3.1 Three-Dimensional Printing Capacity

Capacity planning is the process of determining the equipment capacity required by a manufacturer to meet future demand for its products [1]. 3D printing capacity planning is to select 3D printers and determine the number of each printer to be acquired. However, 3D printing capacity planning is subject to the following challenges. First, although a 3D printer is most suitable for fabricating specific types of 3D objects (or parts), it theoretically can print almost all types of 3D objects, making the selection of an optimal 3D printer a difficult task [2, 3]. In addition, 3D printers are usually applied for the rapid prototyping of new products. However, it is difficult to estimate the demand for such products that are still under development.

In 3D printing capacity planning, the first task is to determine the type of each 3D printer. Choosing a metal printer or a plastic printer depends on the types of 3D objects to be printed. A plastic printer applies printing technologies such as stereolithography (SLA), selective laser sintering (SLS), and material jetting (MJ), while a metal printer applies printing technologies including selective laser melting (SLM), direct metal laser sintering (DMLS), metal binder jetting, and directed energy deposition (DED) [4, 5]. The second task is to determine the scale of each 3D printer. While small 3D printers are often inadequate, a manufacturer tends to acquire a large 3D printer with high printing speed, high resolution, and vast build volume or area. However, a large 3D printer may not be frequently and efficiently utilized. In addition, more powerful 3D printers have been continuously launched at cheaper prices, making the previous decision of acquiring a large 3D printer a bad one. For these reasons, if multiple 3D printers are to be acquired, a capacity planner can minimize the risk by selecting a set of printers with maximal diversity, so as to respond to unexpected or changing demand. For example, Janssen et al. [6] asserted that a 3D printing service provider must fulfill orders with various types of 3D printers, not just plastic 3D printers. Dong et al. [7] compared a traditional method to a 3D printing method. The comparison results showed that the traditional method uses more versatile machines,

© The Author(s), under exclusive license to Springer Nature Switzerland AG 2020
T.-C. Chen, *3D Printing and Ubiquitous Manufacturing*,
SpringerBriefs in Applied Sciences and Technology,
https://doi.org/10.1007/978-3-030-49150-5_3

more complex operating systems, and a more extensively trained workforce than the 3D printing method. By contrast, the 3D printing method delivers higher product variety than the traditional method. Such an advantage can be strengthened by the use of heterogeneous and diverse 3D printers.

A capacity planner usually has various concerns in choosing a suitable 3D printer. To address this, a multiple-criteria decision-making (MCDM) method is applicable. This chapter first describes the application of an analytic hierarchy process (AHP) approach to select the most suitable 3D printer for a manufacturer. Subsequently, by discovering the various viewpoints held by a capacity planner, a decomposition AHP approach is applied to select diverse 3D printers that suit the requirements of the capacity planner.

3.2   Selecting the Most Suitable Three-Dimensional Printer: An Analytic Hierarchy Process Approach

According to Hoffman [8] and Garrett [9], the following factors must be considered before choosing a suitable 3D printer:

- budget,
- user's experience,
- objects to be printed,
- the sizes of objects,
- materials to print with,
- the resolution of objects,
- the requirement for multiple colors,
- the surface for objects to build on,
- the necessity for a closed frame,
- connection to the printer,
- the required software,
- extended warranty, and
- customer service.

In addition, the following measures can be taken for assisting the making of such a choice:

- low price,
- high rating,
- supporting many types of applications,
- vast build volume,
- able to print many colors,
- high resolution,
- supporting many types of materials,
- time-saving, and
- material-saving.

In the AHP approach, a capacity planner compares the relative weight of a factor over that of another using linguistic terms such as "as equal as," "weakly more important than," "strongly more important than," "very strongly more important than," and "absolutely more important than." These linguistic terms are usually mapped to a value within [1, 9, 10]:

L_1: "As equal as" $= 1$
L_2: "Weakly more important than" $= 3$
L_3: "Strongly more important than" $= 5$
L_4: "Very strongly more important than" $= 7$
L_5: "Absolutely more important than" $= 9$

For example, if "time-saving" is strongly more important than "high resolution," then the relative weight of "time-saving" over "a high resolution" is L_3, meaning that the importance of "time-saving" is five times that of "high resolution." If the relative weight is between two successive linguistic terms, values such as 2, 4, 6, and 8 can be chosen. For example, if "material-saving" is strongly more important than or very strongly more important than "vast build volume," then the relative importance of the former over the latter is six times.

Based on pairwise comparison results, the pairwise comparison matrix (or judgment matrix) $\mathbf{A}_{n \times n} = [a_{ij}]$ is constructed:

$$a_{ji} = 1/a_{ij} \tag{3.1}$$

$$a_{ii} = 1 \tag{3.2}$$

The eigenvalue and eigenvector of $\mathbf{A}$, indicated with λ and $\mathbf{x}$ respectively, satisfy

$$det(\mathbf{A} - \lambda \mathbf{I}) = 0 \tag{3.3}$$

and

$$(\mathbf{A} - \lambda \mathbf{I})\mathbf{x} = 0 \tag{3.4}$$

There are n eigenvalues. The maximal eigenvalue is indicated with $\lambda_{\max}$. Based on the value of $\lambda_{\max}$, the consistency among pairwise comparison results can be evaluated using the following indexes [10]:

$$\text{Consistency index (CI): } CI = \frac{\lambda_{\max} - n}{n - 1} \tag{3.5}$$

$$\text{Consistency ratio(CR): } CR = \frac{CI}{RI} \tag{3.6}$$

where RI is the random consistency index [10] (Table 3.1). CR should be less than 0.1 for a small AHP problem. When the size of a judgment matrix is large, the requirement

Table 3.1 Random consistency index

n	1	2	3	4	5	6	7	8	9	10
RI	0	0	0.58	0.9	1.12	1.24	1.32	1.41	1.45	1.49

for CR can be relaxed to less than 0.3. Otherwise, the capacity planner needs to adjust pairwise comparison results or resorts to a fuzzy AHP (FAHP) approach instead.

In addition, the (absolute) importance (or weight) of each factor is derived as

$$w_i = \frac{x_i}{\sum_{j=1}^{n} x_j} \tag{3.7}$$

Theorem 3.1

$$w_i = \frac{\sum_{j=1}^{n} a_{ij}^{\infty}}{\sum_{l=1}^{n} \sum_{j=1}^{n} a_{lj}^{\infty}} \tag{3.8}$$

$$\lambda_{\max} = \frac{1}{n} \sum_{i=1}^{n} \left(\frac{\sum_{j=1}^{n} (a_{ij} x_j)}{x_i} \right) \tag{3.9}$$

Subsequently, the performance of 3D printer k in optimizing factor i is evaluated. The result is indicated with p_{ki}. Then, the overall performance of the 3D printer is derived using the weighted sum as

$$P_k = \sum_{i=1}^{n} w_i p_{ki} \tag{3.10}$$

The most suitable 3D printer is the one that achieves the highest overall performance. An example is given below.

Example 3.1 The AHP approach was applied to select a suitable 3D printer for a manufacturer that considered the following factors:

- low price,
- high rating,
- supporting many types of applications,
- vast build volume, and
- high resolution.

The AHP problem is illustrated in Fig. 3.1. To this end, a capacity planner first compared the relative weights of these factors with linguistic terms. The results are summarized in Table 3.2.

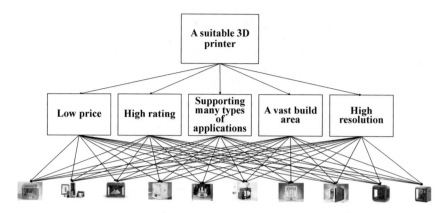

Fig. 3.1 The AHP problem

Table 3.2 Results of pairwise comparisons

Factor #1	Factor #2	Relative weight of factor #1 over factor #2
Low price	High rating	Weakly more important than
Low price	Vast build volume	Strongly or very strongly more important than
Supporting many types of applications	Low price	As equal as
Low price	High resolution	Strongly more important than
High rating	Vast build volume	Very strongly more important than
Supporting many types of applications	High rating	Weakly or strongly more important than
High rating	High resolution	Strongly or very strongly more important than
Supporting many types of applications	Vast build volume	As equal as
Vast build volume	High resolution	As equal as
Supporting many types of applications	High resolution	Weakly or strongly more important than

Based on Table 3.2, the following judgment matrix was constructed:

$$\mathbf{A} = \begin{bmatrix} 1 & 3 & 6 & 1/1 & 5 \\ 1/3 & 1 & 7 & 1/4 & 6 \\ 1/6 & 1/7 & 1 & 1/1 & 1 \\ 1 & 4 & 1 & 1 & 4 \\ 1/5 & 1/6 & 1/1 & 1/4 & 1 \end{bmatrix}$$

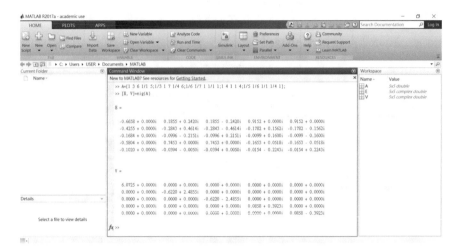

Fig. 3.2 Applying MATLAB R2017a to derive the eigenvalue and eigenvector of A

MATLAB R2017a was applied to derive the eigenvalue and eigenvector of **A**, as illustrated in Fig. 3.2.

The consistency ratio of **A** was evaluated according to Eq. (3.6) as

$$CR(\mathbf{A}) = 0.239$$

which was slightly inconsistent. In addition, the weight of each factor was derived according to Eq. (3.7) as

$$[x_i] = [0.34\ 0.22\ 0.09\ 0.30\ 0.05]^T$$

Among the five factors, price and resolution were the-lower-the-better performance measures, while rating, build volume, and the number of supported application types were the-higher-the-better performance measures. The performances in optimizing these factors were evaluated as follows:

$p_{k1} = $ 1 – (price – the minimal price)/(the maximal price – the minimal price)

$p_{k2} = $ (rating – 1)/(5 – 1)

$p_{k3} = $ (the number of supported application types – the minimal number of supported application types)/(the maximal number of supported application types – the minimal number of supported application types)

$p_{k4} = $ (build volume – the minimal build volume)/(the maximal build volume – the minimal build volume)

$p_{k5} = $ 1 – (resolution – the minimal resolution)/(the maximal resolution – the minimal resolution)

The ten 3D printers recommended by Hoffmann [8], which data are summarized in Table 3.3, were compared.

Table 3.3 Data of ten 3D printers

No.	Printer	Price (US$)	Rating	No. of supported application types	Build volume (inch^3)	Resolution (micron)
1	Dremel DigiLab 3D45 3D Printer	1799	4.5	3	402	50
2	Formlabs Form 2	3500	4	4	224	25
3	MakerBot Replicator+	2799	4	3	573	100
4	Ultimaker S5	5995	4	2	1457	60
5	Dremel DigiLab 3D40 Flex 3D Printer	1034	4	1	402	100
6	LulzBot Mini 2	1485	4	3	282	50
7	Ultimaker 3	4295	4	3	564	60
8	XYZprinting da Vinci Jr. 2.0 Mix	400	4	2	205	200
9	FlashForge Finder 3D Printer	299	3.5	1	166	100
10	Monoprice Voxel 3D Printer	400	3	1	329	60

The evaluated performances are summarized in Table 3.4.

The overall performances of the 3D printers were calculated according to Eq. (3.10). The results are summarized in Table 3.5. The most suitable 3D printer was Dremel DigiLab 3D45 3D Printer, and the second suitable was Ultimaker 3. It

Table 3.4 Evaluated performances

k	Printer	p_{k1}	p_{k2}	p_{k3}	p_{k4}	p_{k5}
1	Dremel DigiLab 3D45 3D Printer	0.737	0.875	0.667	0.183	0.857
2	Formlabs Form 2	0.438	0.750	1.000	0.045	1.000
3	MakerBot Replicator+	0.561	0.750	0.667	0.315	0.571
4	Ultimaker S5	0.000	0.750	0.333	1.000	0.800
5	Dremel DigiLab 3D40 Flex 3D Printer	0.871	0.750	0.000	0.183	0.571
6	LulzBot Mini 2	0.792	0.750	0.667	0.090	0.857
7	Ultimaker 3	0.298	0.750	0.667	0.308	0.800
8	XYZprinting da Vinci Jr. 2.0 Mix	0.982	0.750	0.333	0.030	0.000
9	FlashForge Finder 3D Printer	1.000	0.625	0.000	0.000	0.571
10	Monoprice Voxel 3D Printer	0.982	0.500	0.000	0.126	0.800

Table 3.5 Overall performances of the 3D printers

k	P_k
1	0.602
2	0.467
3	0.539
4	0.534
5	0.548
6	0.566
7	0.458
8	0.539
9	0.510
10	0.526

should be noted that this result was based on the subjective belief of the capacity planner and not absolute.

As a result, if the manufacturer needed to acquire a single 3D printer, the best choice was Dremel DigiLab 3D45 3D Printer. If two 3D printers were to be acquired, both Dremel DigiLab 3D45 3D Printer and LulzBot Mini 2 could be chosen. However, the two 3D printers were quite similar: both were weak in optimizing build volume resolution but strong in optimizing the others. In sum, the two 3D printers did not compensate for each other, which engendered the difficulty in responding to unexpected demand in the future. To solve this problem, a decomposition AHP approach is proposed in the next section [11].

3.3 Selecting Diversified Three-Dimensional Printers: The Decomposition Analytic Hierarchy Process Approach

To build up diversified 3D printing capacity, a manufacturer has to specify the requirements for different types of 3D printers, which is not an easy task. As an alternative, a decomposition AHP approach, extended from the previous AHP approach, is applied to select diversified 3D printers.

In the decomposition AHP approach, the judgement matrix $\mathbf{A}$ is decomposed into several subjudgement matrices $\{\mathbf{A}(l) \mid l = 1 \sim L\}$, to which the arithmetic average operator is applied:

$$\mathbf{A} \overset{def}{=} \frac{\sum_{l=1}^{L} \mathbf{A}(l)}{L} \tag{3.11}$$

All subjudgement matrices meet the basic requirements:

$$det\left(\mathbf{A}(l) - \lambda(l)\mathbf{I}\right) = 0 \tag{3.12}$$

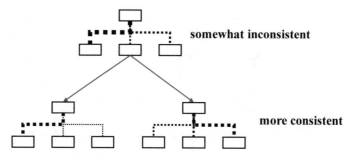

somewhat inconsistent

more consistent

Fig. 3.3 Decomposing the judgment matrix into several subjudgment matrices

$$(\mathbf{A}(l) - \lambda(l)\mathbf{I})\mathbf{x}(l) = 0 \tag{3.13}$$

It is expected that by considering a single viewpoint at a time, each subjudgment matrix will be more consistent than the original judgment matrix.

$$CR(\mathbf{A}(l)) \leq CR(\mathbf{A}) \, ; l = 1 \sim L \tag{3.14}$$

This is illustrated in Fig. 3.3.

However, the possible combinations of subjudgment matrices are numerous. In the decomposition AHP approach, an optimal combination of subjudgment matrices is chosen such that these matrices are as far from each other as possible.

$$\text{Max } Z_1 = \sum_{l=1}^{L-1} \sum_{m=l+1}^{L} d(\mathbf{A}(l), \ \mathbf{A}(m)) \tag{3.15}$$

where $d()$ is the fuzzy distance function. The objective is to consider viewpoints that are as diverse as possible. As a result, the following mixed-integer nonlinear programming (MINLP) model is formulated and optimized:

(MINLP Model)

$$\text{Max } Z_1 = \sum_{l=1}^{L-1} \sum_{m=l+1}^{L} d(\mathbf{A}(l), \ \mathbf{A}(m)) \tag{3.16}$$

subject to

$$\mathbf{A} \overset{def}{=} \frac{\sum_{l=1}^{L} \mathbf{A}(l)}{L} \tag{3.17}$$

$$CR(\mathbf{A}(l)) \leq CR(\mathbf{A}); l = 1 \sim L \tag{3.18}$$

$$det\,(\mathbf{A}(l) - \lambda(l)\mathbf{I}) = 0; l = 1 \sim L \tag{3.19}$$

$$(\mathbf{A}(l) - \lambda(l)\mathbf{I})\mathbf{x}(l) = 0; l = 1 \sim L \tag{3.20}$$

$$a_{ii}(l) = 1; i = 1 \sim n; l = 1 \sim L \tag{3.21}$$

$$a_{ij}(l) \in [1,\ 9] \forall a_{ij}(l) \geq 1; i, j = 1 \sim n; i \neq j; l = 1 \sim L \tag{3.22}$$

$$a_{ji}(l) = 1/a_{ij}(l) \forall a_{ij}(l) \geq 1; i, j = 1 \sim n; i \neq j; l = 1 \sim L \tag{3.23}$$

Constraints (3.19) to (3.23) are the requirements for a judgment (or subjudgment) matrix. The Frobenius distance [12] can be used as $d()$:

$$d(\mathbf{A}(l),\ \mathbf{A}(m)) = \sqrt{\sum_{i=1}^{n} \sum_{j=1}^{n} (a_{ij}(l) - a_{ij}(m))^2} \tag{3.24}$$

Example 3.2 In the previous example, the CR of **A** was 0.239, which was somewhat inconsistent. Therefore, the capacity planner decided to decompose the judgment matrix into two subjudgment matrices to discover his multiple viewpoints. Toward this end, the MINLP model was built and solved with an enumeration procedure by using MATLAB R2017a, as presented in Fig. 3.4.

The optimal solution was obtained as follows:

$$\mathbf{A}^*(1) = \begin{bmatrix} 1 & 1 & 3 & 1/1 & 2 \\ 1/1 & 1 & 9 & 1/3 & 9 \\ 1/3 & 1/9 & 1 & 1/1 & 1 \\ 1 & 3 & 1 & 1 & 5 \\ 1/2 & 1/9 & 1/1 & 1/5 & 1 \end{bmatrix}; \ \mathbf{A}^*(2) = \begin{bmatrix} 1 & 5 & 9 & 1/1 & 8 \\ 1/5 & 1 & 5 & 1/5 & 3 \\ 1/9 & 1/5 & 1 & 1/1 & 1 \\ 1 & 5 & 1 & 1 & 3 \\ 1/8 & 1/3 & 1/1 & 1/3 & 1 \end{bmatrix}$$

The solution provided the following parameter values: $Z_1^* = 12.203$, $CR(\mathbf{A}^*(1)) = 0.239$, and $CR(\mathbf{A}^*(2)) = 0.226$. Obviously, the subjudgment matrices were more consistent than the original judgment matrix. The weights of factors determined by the two subjudgment matrices were {0.206, 0.331, 0.094, 0.313, 0.059} and {0.408, 0.135, 0.077, 0.262, 0.052}, respectively. The weights derived from either subjudgment matrix were used to evaluate the overall performance of each 3D printer according to the weighted sum. The results are summarized in Table 3.6.

The top-performing 3D printers from the two viewpoints were Ultimaker S5 and Dremel DigiLab 3D45 3D Printer, respectively. The performances of the two 3D printers were quite different in optimizing price and build volume, thereby ensuring a diversified choice of 3D printers.

```
A=[1 3 6 1/1 5; 1/3 1 7 1/4 6; 1/6 1/7 1 1/1 1; 1 4 1 1 4; 1/5 1/6 1/1 1/4 1];
[E V]=eig(A);
CR=((V(1,1)-5)/(5-1))/1.12;
A1best=zeros(5,5);
A2best=zeros(5,5);
CR1best=0;
CR2best=0;
E1best=zeros(5,5);
E2best=zeros(5,5);
distbest=0;

for i12=1:5
for i13=3:9
for i15=1:9
for i23=5:9
for i25=3:9
for i35=1:1
for i41=1:1
for i42=1:7
for i43=1:1
for i45=1:7
A1=A;
A2=A;
A1(1,2)=i12; A1(2,1)=1/A1(1,2);
A1(1,3)=i13; A1(3,1)=1/A1(1,3);
A1(1,5)=i15; A1(5,1)=1/A1(1,5);
A1(2,3)=i23; A1(3,2)=1/A1(2,3);
A1(2,5)=i25; A1(5,2)=1/A1(2,5);
A1(3,5)=i35; A1(5,3)=1/A1(3,5);
A1(4,1)=i41; A1(1,4)=1/A1(4,1);
A1(4,2)=i42; A1(2,4)=1/A1(4,2);
A1(4,3)=i43; A1(3,4)=1/A1(4,3);
A1(4,5)=i45; A1(5,4)=1/A1(4,5);
A2(1,2)=2*A(1,2)-A1(1,2); A2(2,1)=1/A2(1,2);
A2(1,3)=2*A(1,3)-A1(1,3); A2(3,1)=1/A2(1,3);
A2(1,5)=2*A(1,5)-A1(1,5); A2(5,1)=1/A2(1,5);
A2(2,3)=2*A(2,3)-A1(2,3); A2(3,2)=1/A2(2,3);
A2(2,5)=2*A(2,5)-A1(2,5); A2(5,2)=1/A2(2,5);
A2(3,5)=2*A(3,5)-A1(3,5); A2(5,3)=1/A2(3,5);
A2(4,1)=2*A(4,1)-A1(4,1); A2(1,4)=1/A2(4,1);
A2(4,2)=2*A(4,2)-A1(4,2); A2(2,4)=1/A2(4,2);
A2(4,3)=2*A(4,3)-A1(4,3); A2(3,4)=1/A2(4,3);
A2(4,5)=2*A(4,5)-A1(4,5); A2(5,4)=1/A2(4,5);
[E1,V1]=eig(A1);
[E2,V2]=eig(A2);
CR1=((V1(1,1)-5)/(5-1))/1.12;
CR2=((V2(1,1)-5)/(5-1))/1.12;
if CR1<=CR & CR2<=CR
dist=sqrt(sum(sum((A1-A2).^2)));
if dist>distbest
distbest=dist;
A1best=A1;
A2best=A2;
CR1best=CR1;
CR2best=CR2;
E1best=E1;
E2best=E2;
end
end
end
end
end
end
end
end
end
end
end
end
end
```

Fig. 3.4 The enumeration procedure

Table 3.6 The evaluation results

p	$P_k(1)$	$P_k(2)$
1	0.610	0.561
2	0.504	0.420
3	0.557	0.493
4	0.638	0.430
5	0.517	0.533
6	0.551	0.543
7	0.514	0.396
8	0.490	0.534
9	0.445	0.521
10	0.453	0.542

3.4 A Fuzzy Analytic Hierarchy Process Approach for Selecting the Most Suitable Three-Dimensional Printer

In the FAHP approach, the linguistic terms for a pairwise comparison are usually mapped to triangular fuzzy numbers (TFNs) within [1, 9] (see Fig. 3.5) [13]:

$\tilde{L}_1$: "As equal as" $= (1, 1, 3)$
$\tilde{L}_2$: "Weakly more important than" $= (1, 3, 5)$
$\tilde{L}_3$: "Strongly more important than" $= (3, 5, 7)$
$\tilde{L}_4$: "Very strongly more important than" $= (5, 7, 9)$
$\tilde{L}_5$: "Absolutely more important than" $= (7, 9, 9)$

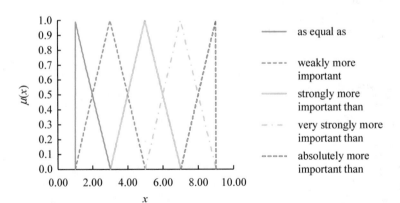

Fig. 3.5 Linguistic terms and the corresponding TFNs used in the FAHP approach

so as to represent the subjective beliefs of more capacity planners. If the relative weight is between two successive linguistic terms, values such as (1, 2, 4), (2, 4, 6), (4, 6, 8), and (6, 8, 9) can be chosen.

Based on the fuzzy pairwise comparison results, the fuzzy judgment matrix $\tilde{\mathbf{A}}_{n \times n} = [\tilde{a}_{ij}]$ is constructed. The fuzzy eigenvalue and eigenvector of $\tilde{\mathbf{A}}$, indicated with $\tilde{\lambda}$ and $\tilde{\mathbf{x}}$ respectively, satisfy

$$det\,(\tilde{\mathbf{A}}(-)\tilde{\lambda}\mathbf{I}) = 0 \qquad (3.25)$$

and

$$(\tilde{\mathbf{A}}(-)\tilde{\lambda}\mathbf{I})(\times)\tilde{\mathbf{x}} = 0 \qquad (3.26)$$

where $(-)$ and $(\times)$ denote fuzzy subtraction and multiplication, respectively.

Various methods have been proposed to solve a FAHP problem by deriving or approximating the eigenvalue and eigenvector of the fuzzy pairwise comparison matrix, for example, the fuzzy geometric mean (FGM) method proposed by Buckley [14], the fuzzy extent analysis (FEA) method proposed by Chang [15], the method based on α-cut operations (ACO) [16, 17], the fuzzy prioritization method (FPM) [18], and others. The first three methods are more prevalent than the last method. Each method has its variants.

The fuzzy geometric mean (FGM) method is applied to approximate the value of the fuzzy weight of each factor $\tilde{w}_i$ as

$$\tilde{w}_i \cong \frac{\sqrt[n]{\prod_{j=1}^{n} \tilde{a}_{ij}}}{\sum_{k=1}^{n} \sqrt[n]{\prod_{j=1}^{n} \tilde{a}_{kj}}} \qquad (3.27)$$

which can be expanded as

$$w_{i1} \cong \frac{1}{1 + \sum_{k \neq i} \frac{\sqrt[n]{\prod_{j=1}^{n} a_{kj3}}}{\sqrt[n]{\prod_{j=1}^{n} a_{ij1}}}} \qquad (3.28)$$

$$w_{i2} \cong \frac{1}{1 + \sum_{k \neq i} \frac{\sqrt[n]{\prod_{j=1}^{n} a_{kj2}}}{\sqrt[n]{\prod_{j=1}^{n} a_{ij2}}}} \qquad (3.29)$$

$$w_{i3} \cong \frac{1}{1 + \sum_{k \neq i} \frac{\sqrt[n]{\prod_{j=1}^{n} a_{kj1}}}{\sqrt[n]{\prod_{j=1}^{n} a_{ij3}}}} \qquad (3.30)$$

In addition, the fuzzy maximal eigenvalue $\tilde{\lambda}_{\max}$ can be estimated as

$$\tilde{\lambda}_{max} \cong \frac{1}{n} \sum_{i=1}^{n} \frac{\sum_{j=1}^{n} (\tilde{a}_{ij}(\times)\tilde{w}_j)}{\tilde{w}_i} \tag{3.31}$$

which is equivalent to

$$\lambda_{max,1} \cong 1 + \frac{1}{n} \sum_{i=1}^{n} \sum_{j \neq i} \frac{a_{ij1} w_{j1}}{w_{i3}} \tag{3.32}$$

$$\lambda_{max,2} \cong 1 + \frac{1}{n} \sum_{i=1}^{n} \sum_{j \neq i} \frac{a_{ij2} w_{j2}}{w_{i2}} \tag{3.33}$$

$$\lambda_{max,3} \cong 1 + \frac{1}{n} \sum_{i=1}^{n} \sum_{j \neq i} \frac{a_{ij3} w_{j3}}{w_{i1}} \tag{3.34}$$

Based on the value of $\tilde{\lambda}_{max}$, the consistency among fuzzy pairwise comparison results can be evaluated as [19]:

Fuzzy consistency index:

$$\widetilde{CI} = \frac{\tilde{\lambda}_{max} - n}{n - 1} \tag{3.35}$$

Fuzzy consistency ratio:

$$\widetilde{CR} = \frac{\widetilde{CI}}{RI} \tag{3.36}$$

$\widetilde{CR}$ should be less than 0.1 for a small FAHP problem:

- If $CR_3 \leq 0.1$, pairwise comparison results are completely consistent.
- If $CR_1 > 0.1$, pairwise comparison results are completely inconsistent.
- Otherwise, pairwise comparison results are somewhat consistent (or inconsistent).

When the size of the judgment matrix is large, the requirement for $\widetilde{CR}$ can be relaxed to less than 0.3.

The most suitable 3D printer is the one that achieves the highest overall performance in terms of the fuzzy weighted sum:

$$\tilde{P}_k = (P_{k1}, \ P_{k2}, \ P_{k3})$$
$$= \sum_{i=1}^{n} (\tilde{w}_i(\times)\tilde{p}_{ki})$$
$$= (w_{i1} p_{ki1}, \ w_{i2} p_{ki2}, \ w_{i3} p_{ki3}) \tag{3.37}$$

Table 3.7 Fuzzy weights of factors

i	$\tilde{w}_i$
1	(0.170, 0.388, 0.521)
2	(0.110, 0.203, 0.388)
3	(0.037, 0.075, 0.152)
4	(0.172, 0.275, 0.529)
5	(0.027, 0.061, 0.119)

where $\tilde{p}_{ki}$ is the fuzzy performance of the k-th 3D printer in optimizing the i-th factor. $\tilde{P}_k$ can be defuzzified using the center-of-gravity (COG) method for an absolute ranking [20]:

$$D(\tilde{P}_k) = \frac{P_{k1} + P_{k2} + P_{k3}}{3} \tag{3.38}$$

Example 3.3 In Example 3.1, the FAHP approach was applied instead to select a suitable 3D printer for the manufacturer. Based on Table 3.2, the following fuzzy judgment matrix was constructed:

$$\tilde{A} = \begin{bmatrix} 1 & (1, 3, 5) & (4, 6, 8) & 1/(1, 1, 3) & (3, 5, 7) \\ 1/(1, 3, 5) & 1 & (5, 7, 9) & 1/(2, 4, 6) & (4, 6, 8) \\ 1/(4, 6, 8) & 1/(5, 7, 9) & 1 & 1/(1, 1, 3) & (1, 1, 3) \\ (1, 1, 3) & (2, 4, 6) & (1, 1, 3) & 1 & (2, 4, 6) \\ 1/(3, 5, 7) & 1/(4, 6, 8) & 1/(1, 1, 3) & 1/(2, 4, 6) & 1 \end{bmatrix}$$

The fuzzy weight of each factor was derived according to Eq. (3.27). The results are summarized in Table 3.7. Taking the weight of "vast build volume" as an example. It could be as small as 0.172, or up to 0.529, giving the capacity planer much flexibility in comparing these factors, which is an advantage of the FAHP approach over the AHP approach. The fuzzy consistency ratio of $\tilde{A}$ was evaluated according to Eq. (3.31) as

$$\widetilde{CR}(\tilde{A}) = (-0.713, \ 0.224, \ 5.320)$$

which was somewhat inconsistent.

The overall performances of the 3D printers were evaluated according to Eq. (3.37), and the results are summarized in Table 3.8. The overall performance was then defuzzified using the COG method. The most suitable 3D printer was Dremel DigiLab 3D45 3D Printer, and the second suitable was LulzBot Mini 2. It should be noted that the results were somewhat different from those obtained using the AHP method.

As a result, when the manufacturer needed to acquire a single 3D printer, the best choice was Dremel DigiLab 3D45 3D Printer. If two 3D printers were to be acquired, both Dremel DigiLab 3D45 3D Printer and LulzBot Mini 2 could be chosen.

Table 3.8 The overall performances of the 3D printers

k	$\tilde{P}_k$	$D(\tilde{P}_k)$
1	(0.301, 0.615, 1.024)	0.647
2	(0.229, 0.469, 0.815)	0.504
3	(0.272, 0.540, 0.920)	0.577
4	(0.289, 0.500, 0.966)	0.585
5	(0.277, 0.574, 0.910)	0.587
6	(0.281, 0.585, 0.955)	0.607
7	(0.233, 0.450, 0.806)	0.496
8	(0.267, 0.566, 0.869)	0.567
9	(0.254, 0.549, 0.832)	0.545
10	(0.265, 0.565, 0.868)	0.566

References

1. W.J. Stevenson, M. Hojati, J. Cao J, *Operations Management* (McGraw-Hill/Irwin, Boston, 2007)
2. D. Jamie, How to choose a 3D printer: experts give their advice (2018). https://www.3dnati ves.com/en/how-to-choose-a-3d-printer-260120184/
3. Pick3DPrinter.com, What is 3D printing? A quick guide for beginners (2019). https://pick3d printer.com/what-is-3d-printing/
4. D Systems, Types of plastic 3D printing (2020). https://www.3dsystems.com/advantages-of-plastic-3D-printing
5. A.B. Varotsis, Introduction to metal 3D printing (2020). https://www.3dhubs.com/knowledge-base/introduction-metal-3d-printing/
6. R. Janssen, I. Blankers, E. Moolenburgh, B. Posthumus, TNO: the impact of 3-D printing on supply chain management. TNO **28**, 24 (2014)
7. L. Dong, D. Shi, F. Zhang, 3D printing vs. traditional flexible technology: implications for manufacturing strategy. Available at SSRN 2847731 (2016)
8. T. Hoffman, The best 3D printers for 2019 (2019). https://www.pcmag.com/roundup/328263/the-best-3d-printers
9. C. Garrett, How to choose the Right 3D printer for you (2016). https://makerhacks.com/cho ose-3d-printer/
10. T.L. Saaty, Decision making with the analytic hierarchy process. Int. J. Ser. Sci. **1**(1), 83–98 (2008)
11. Y.-C. Lin, T. Chen, A multibelief analytic hierarchy process approach for diversifying product designs: smart backpack design as an example. J. Eng. Manuf. (2019)
12. G.H. Golub, C.F. Van Loan, *Matrix Computations* (Johns Hopkins, Baltimore, 1996)
13. G. Zheng, N. Zhu, Z. Tian, Y. Chen, B. Sun, Application of a trapezoidal fuzzy AHP method for work safety evaluation and early warning rating of hot and humid environments. Saf. Sci. **50**(2), 228–239 (2012)
14. J.J. Buckley, Fuzzy hierarchical analysis. Fuzzy Sets Syst. **17**(3), 233–247 (1985)
15. D.Y. Chang, Applications of the extent analysis method on fuzzy AHP. Eur. J. Oper. Res. **95**(3), 649–655 (1996)
16. C.H. Cheng, D.L. Mon, Evaluating weapon system by analytical hierarchy process based on fuzzy scales. Fuzzy Sets Syst. **63**(1), 1–10 (1994)
17. T. Chen, Y.-C. Lin, M.-C. Chiu, Approximating alpha-cut operations approach for effective and efficient fuzzy analytic hierarchy process analysis. Appl. Soft Comput. **85**, 105855 (2019)

18. L. Wang, J. Chu, J. Wu, Selection of optimum maintenance strategies based on a fuzzy analytic hierarchy process. Int. J. Prod. Econ. **107**(1), 151–163 (2007)
19. Y.C. Wang, T.C.T. Chen, A partial-consensus posterior-aggregation FAHP method—supplier selection problem as an example. Mathematics **7**(2), 179 (2019)
20. E. Van Broekhoven, B. De Baets, Fast and accurate center of gravity defuzzification of fuzzy system outputs defined on trapezoidal fuzzy partitions. Fuzzy Sets Syst. **157**(7), 904–918 (2006)

Chapter 4
Capacity Planning for a Ubiquitous Manufacturing System Based on Three-Dimensional Printing

4.1 Capacity and Production Planning Procedure

The capacity and production planning of a three-dimensional (3D) printing-based ubiquitous manufacturing (UM) system comprises the following steps:

(1) 3D printing capacity planning for a manufacturer.
(2) Capacity planning for a UM system.
(3) Production planning for a UM system.

as illustrated in Fig. 4.1. The first step has been described in Chap. 3. This chapter is dedicated to the second step. The third step will be discussed in Chap. 5.

Printing a 3D object usually takes hours to tens of hours, which is an obstacle to the application of 3D printing for mass production. Consequently, to mass produce a product, a factory must purchase a number of 3D printers, or resort to the service of a UM system. The capacity of a UM system composed of 3D printers, computer numerical control (CNC) machines, and/or robots is easy to expand or shrink [1]. A UM system composed of distributed 3D printing facilities is the focus of this chapter. When an order is received by such a UM system, it is split among several 3D printing facilities that are available, have fast printing speeds, and close to the customer [2], which involves both production and transportation planning. However, a 3D printing-based UM system can be further benefited by addressing the following concerns [3, 4]:

(1) The quality of products can be improved to attract and retain customers.
(2) The costs of fabricating products can be reduced to be more competitive.
(3) A long-term relationship with each 3D printing facility can be fostered for the UM system to be sustainable.

as illustrated in Fig. 4.2.

© The Author(s), under exclusive license to Springer Nature Switzerland AG 2020
T.-C. Chen, *3D Printing and Ubiquitous Manufacturing*,
SpringerBriefs in Applied Sciences and Technology,
https://doi.org/10.1007/978-3-030-49150-5_4

Fig. 4.1 The procedure for
the capacity and production
planning of a 3D
printing-based UM system

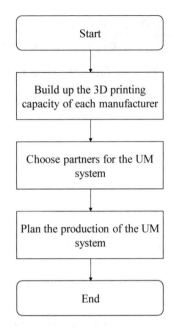

Fig. 4.2 Considerations in
planning the capacity of a 3D
printing-based UM system

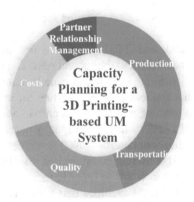

A 3D printing facility is not always available for participating in a 3D printing-based UM system. In addition, a 3D printing facility may serve multiple UM systems at the same time, as illustrated in Fig. 4.3. As a result, 3D printing facilities participating in a 3D printing-based UM system may change. Therefore, capacity planning for a 3D printing-based UM system is the dynamic process of assessing and choosing 3D printing facilities to be incorporated in the 3D printing-based UM system. To fulfill this task, the application of analytic hierarchy process (AHP) and the technique for order preference by similarity to the ideal solution (TOPSIS) is described in this chapter.

Fig. 4.3 A 3D printing
facility may serve multiple
UM systems at the same time

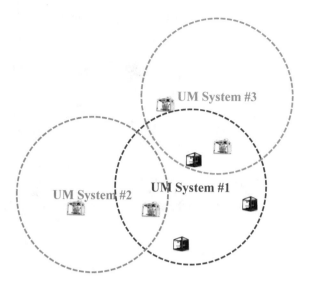

4.2   System Architecture of a Three-Dimensional Printing-Based Ubiquitous Manufacturing System

The system architecture of a 3D printing-based UM system is composed of four parts: customers, the UM service provider, 3D printing facilities, and a transportation service provider, as illustrated in Fig. 4.4. The UM service provider comprises the system administrator, the system server, the system database, and the reasoning module. In this chapter, the reasoning module is the AHP-TOPSIS approach.

The UM system operates according to the following procedure:

Step 1. A customer places an order online.

Step 2. The UM system administrator searches for available 3D printing facilities in the service region.

Step 3. The system administrator negotiates with 3D printing facilities.

Step 4. The system administrator chooses 3D printing facilities using the AHP-TOPSIS approach.

Step 5. The order is distributed among 3D printing facilities.

Step 6. 3D printing facilities print the assigned pieces.

Step 7. A transportation service provider visits all 3D printing facilities to pick up the printed pieces, or each 3D printing facility delivers the printed pieces by itself.

Step 8. The customer feeds back his/her assessment of the service and product quality.

Step 9. The system database is updated with the completion details and the customer's feedback.

The flowchart in Fig. 4.5 illustrates this procedure.

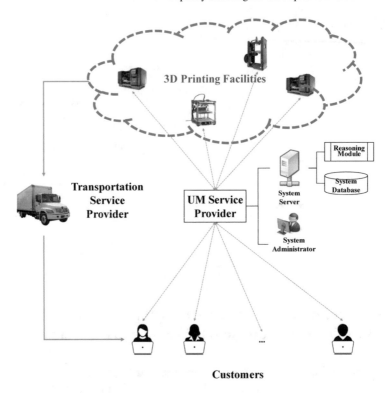

Fig. 4.4   The 3D printing-based UM system architecture

4.3   Assessing and Choosing Three-Dimensional Printing Facilities

4.3.1   Determining the Weights of Criteria

According to [2–4], the following issues must be considered when assessing a 3D printing facility:

- the printing speed,
- the delivery speed, if the printed pieces are delivered by the 3D printing facility itself,
- the relationship between the UM system and the 3D printing facility,
- the quality of 3D items printed by the 3D printing facility, and
- the average cost.

The performances of a 3D printing facility in optimizing these criteria can be evaluated according to the rules in Table 4.1 [5].

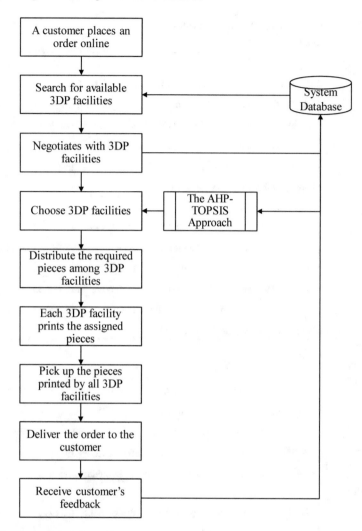

Fig. 4.5 The operational procedure of the UM system

In the AHP-TOPSIS approach, the UM system administrator first applies AHP to compare the relative weight of a factor over that of another using linguistic terms such as "as equal as," "weakly more important than," "strongly more important than," "very strongly more important than," and "absolutely more important than." These linguistic terms are mapped to a value within [1, 9] [6]:

L_1: "As equal as" = 1
L_2: "Weakly more important than" = 3
L_3: "Strongly more important than" = 5
L_4: "Very strongly more important than" = 7
L_5: "Absolutely more important than" = 9

Table 4.1 Rules for assessing a 3D printing facility

Criterion	Rule
The printing speed performance (p_{k1})	$p_{k1}(x_k) =$ $\begin{cases} 0 & \text{if} & 0.1 \cdot \min_r x_r + 0.9 \cdot \max_r x_r \leq x_k \\ 1 & \text{if} & 0.35 \cdot \min_r x_r + 0.65 \cdot \max_r x_r \leq x_k < 0.1 \cdot \min_r x_r + 0.9 \cdot \max_r x_r \\ 2.5 & \text{if} & 0.65 \cdot \min_r x_r + 0.35 \cdot \max_r x_r \leq x_k < 0.35 \cdot \min_r x_r + 0.65 \cdot \max_r x_r \\ 4 & \text{if} & 0.9 \cdot \min_r x_r + 0.1 \cdot \max_r x_r \leq x_k < 0.65 \cdot \min_r x_r + 0.35 \cdot \max_r x_r \\ 5 & \text{if} & \min_r x_r \leq x_k < 0.9 \cdot \min_r x_r + 0.1 \cdot \max_r x_r \end{cases}$ where x_k is the estimated completion time
The delivery speed performance (p_{k2})	$p_{k2}(x_k) =$ $\begin{cases} 0 & \text{if} & 0.1 \cdot \min_r x_r + 0.9 \cdot \max_r x_r \leq x_k \\ 1 & \text{if} & 0.35 \cdot \min_r x_r + 0.65 \cdot \max_r x_r \leq x_k < 0.1 \cdot \min_r x_r + 0.9 \cdot \max_r x_r \\ 2.5 & \text{if} & 0.65 \cdot \min_r x_r + 0.35 \cdot \max_r x_r \leq x_k < 0.35 \cdot \min_r x_r + 0.65 \cdot \max_r x_r \\ 4 & \text{if} & 0.9 \cdot \min_r x_r + 0.1 \cdot \max_r x_r \leq x_k < 0.65 \cdot \min_r x_r + 0.35 \cdot \max_r x_r \\ 5 & \text{if} & \min_r x_r \leq x_k < 0.9 \cdot \min_r x_r + 0.1 \cdot \max_r x_r \end{cases}$ where x_k is the estimated delivery time
The relationship performance (p_{k3})	$p_{k3}(x_k) =$ $\begin{cases} 0 & \text{if} & \min_r x_r \leq x_k < 0.9 \cdot \min_r x_r + 0.1 \cdot \max_r x_r \\ 1 & \text{if} & 0.9 \cdot \min_r x_r + 0.1 \cdot \max_r x_r \leq x_k < 0.65 \cdot \min_r x_r + 0.35 \cdot \max_r x_r \\ 2.5 & \text{if} & 0.65 \cdot \min_r x_r + 0.35 \cdot \max_r x_r \leq x_k < 0.35 \cdot \min_r x_r + 0.65 \cdot \max_r x_r \\ 4 & \text{if} & 0.35 \cdot \min_r x_r + 0.65 \cdot \max_r x_r \leq x_k < 0.1 \cdot \min_r x_r + 0.9 \cdot \max_r x_r \\ 5 & \text{if} & 0.1 \cdot \min_r x_r + 0.9 \cdot \max_r x_r \leq x_k \end{cases}$ where x_k is the number of orders printed within the last month
The product quality performance (p_{q4})	$p_{k4}(x_k) =$ $\begin{cases} 0 & \text{if} & \min_r x_r \leq x_k < 0.9 \cdot \min_r x_r + 0.1 \cdot \max_r x_r \\ 1 & \text{if} & 0.9 \cdot \min_r x_r + 0.1 \cdot \max_r x_r \leq x_k < 0.65 \cdot \min_r x_r + 0.35 \cdot \max_r x_r \\ 2.5 & \text{if} & 0.65 \cdot \min_r x_r + 0.35 \cdot \max_r x_r \leq x_k < 0.35 \cdot \min_r x_r + 0.65 \cdot \max_r x_r \\ 4 & \text{if} & 0.35 \cdot \min_r x_r + 0.65 \cdot \max_r x_r \leq x_k < 0.1 \cdot \min_r x_r + 0.9 \cdot \max_r x_r \\ 5 & \text{if} & 0.1 \cdot \min_r x_r + 0.9 \cdot \max_r x_r \leq x_k \end{cases}$ where x_k is the average product quality

(continued)

Table 4.1 (continued)

Criterion	Rule
The average cost performance (p_{k5})	$p_{k5}(x_k) =$ $\begin{cases} 0 & \text{if} & 0.1 \cdot \min_r x_r + 0.9 \cdot \max_r x_r \leq x_k \\ 1 & \text{if} & 0.35 \cdot \min_r x_r + 0.65 \cdot \max_r x_r \leq x_k < 0.1 \cdot \min_r x_r + 0.9 \cdot \max_r x_r \\ 2.5 & \text{if} & 0.65 \cdot \min_r x_r + 0.35 \cdot \max_r x_r \leq x_k < 0.35 \cdot \min_r x_r + 0.65 \cdot \max_r x_r \\ 4 & \text{if} & 0.9 \cdot \min_r x_r + 0.1 \cdot \max_r x_r \leq x_k < 0.65 \cdot \min_r x_r + 0.35 \cdot \max_r x_r \\ 5 & \text{if} & \min_r x_r \leq x_k < 0.9 \cdot \min_r x_r + 0.1 \cdot \max_r x_r \end{cases}$ where x_k is the average cost

For example, if "the average cost" is weakly more important than "product quality," then the relative weight of "the average cost" over "product quality" is L_2, meaning that the importance of "the average cost" is three times that of "product quality." If the relative weight is between two successive linguistic terms, values such as 2, 4, 6, and 8 can be chosen.

Based on the pairwise comparison results, the pairwise comparison matrix (or judgment matrix) $\mathbf{A}_{n \times n} = [a_{ij}]$ is constructed:

$$a_{ji} = 1/a_{ij} \tag{4.1}$$

$$a_{ii} = 1 \tag{4.2}$$

The eigenvalue and eigenvector of $\mathbf{A}$, indicated with λ and $\mathbf{x}$, respectively, satisfy

$$det(\mathbf{A} - \lambda \mathbf{I}) = 0 \tag{4.3}$$

and

$$(\mathbf{A} - \lambda \mathbf{I})\mathbf{x} = 0 \tag{4.4}$$

The maximal eigenvalue is indicated with λ_{max}. Based on the value of λ_{max}, the consistency among pairwise comparison results can be evaluated in terms of the consistency ratio (CR) [6]:

$$CR = \frac{CI}{RI} \tag{4.5}$$

where

$$CI = \frac{\lambda_{max} - n}{n - 1} \tag{4.6}$$

RI is the random consistency index [6]. *CR* should be less than 0.1 for a small AHP problem. When the size of the judgment matrix is large, the requirement for *CR* can be relaxed to less than 0.3. Otherwise, the UM system administrator needs to adjust the pairwise comparison results. In addition, the (absolute) importance (or weight) of each factor is derived as

$$w_i = \frac{x_i}{\sum_{j=1}^{n} x_j} \tag{4.7}$$

or

$$w_i = \frac{\sum_{j=1}^{n} a_{ij}^{\infty}}{\sum_{l=1}^{n} \sum_{j=1}^{n} a_{lj}^{\infty}} \tag{4.8}$$

4.3.2 Assessing the Overall Performance of a Three-Dimensional Printing Facility

Subsequently, TOPSIS is applied to assess the overall performance of a 3D printing facility. TOPSIS has been extensively applied to multiple-criteria decision-making problems in various fields [7]. Compared to the weighted sum approach, TOPSIS is more sensitive in detecting a minor difference in the overall performance [8]. Basically, TOPSIS can be applied alone, based on the values of weights specified by the decision-maker [7]. Nevertheless, the joint application of AHP and TOPSIS is prevalent. For example, Lin et al. [9] applied the combination of AHP and TOPSIS for the computer-aided design of a product. First, AHP was applied to determine the priorities of the attributes of a product, in order to determine the optimal product design. Subsequently, TOPSIS was applied to compare the performances of various product designs. Wang et al. [10] proposed a fuzzy AHP-TOPSIS method for supplier selection. Büyüközkan and Çifçi [11] applied a similar method to compare the green competencies of suppliers. They also applied the fuzzy decision-making trial and evaluation laboratory model (DEMATEL) to consider the mutual interdependencies among criteria. Junior et al. [12] compared the advantages and disadvantages of fuzzy TOPSIS and fuzzy AHP on supplier selection and concluded that fuzzy TOPSIS performed better with regard to the change of alternatives or criteria, agility, and the number of criteria and alternatives.

First, the weight of each criterion is multiplied to the normalized performance to derive the weighted score:

$$s_{ki} = w_i \rho_{ki} \tag{4.9}$$

where ρ_{ki} is the normalized performance of the *k*-th 3D printing facility in optimizing the *i*-th criterion, which is derived using ideal normalization as

$$\rho_{ki} = \frac{p_{ki}}{\max\limits_{r} p_{ri}} \tag{4.10}$$

or using distributive normalization as

$$\rho_{ki} = \frac{p_{ki}}{\sqrt{\sum_{r} p_{ri}^2}}. \tag{4.11}$$

where p_{ki} is the un-normalized (original) performance. Two reference points, the ideal point $\boldsymbol{\Lambda}^+ = [\Lambda_i^+]$ and the anti-ideal point $\boldsymbol{\Lambda}^- = [\Lambda_i^-]$, are established, respectively, as

$$\Lambda_i^+ = \max_{k} s_{ki} \tag{4.12}$$

$$\Lambda_i^- = \min_{k} s_{ki} \tag{4.13}$$

The distances from a 3D printing facility to the two reference points are measured:

$$d_k^+ = \sqrt{\sum_{i=1}^{n} (\Lambda_i^+ - s_{ki})^2} \tag{4.14}$$

$$d_k^- = \sqrt{\sum_{i=1}^{n} (\Lambda_i^- - s_{ki})^2} \tag{4.15}$$

Finally, both distances are considered in calculating the closeness:

$$C_k = \frac{d_k^-}{d_k^+ + d_k^-} \tag{4.16}$$

A 3D printing facility is chosen if its closeness is higher.

Example 4.1

The AHP-TOPSIS approach was applied to a UM system that is illustrated in Fig. 4.6. The service region of the UM system covered an area of 47.6 km². There were more than ten 3D printing facilities in this region.

At first, to compare the relative weights of criteria, the UM system administrator constructed the following judgment matrix:

Fig. 4.6 A UM system

$$\mathbf{A} = \begin{bmatrix} 1 & 5 & 3 & 3 & 7 \\ 1/5 & 1 & 1/3 & 1/9 & 1/7 \\ 1/3 & 3 & 1 & 1/3 & 1 \\ 1/3 & 9 & 3 & 1 & 7 \\ 1/7 & 7 & 1/1 & 1/7 & 1 \end{bmatrix}$$

MATLAB R2017a was applied to derive the eigenvalue and eigenvector of $\mathbf{A}$, as illustrated in Fig. 4.7.

The consistency ratio of $\mathbf{A}$ was evaluated as

$$CR(\mathbf{A}) = 0.154$$

which was slightly inconsistent but still tolerable. The weights of the five criteria—the printing speed, the delivery speed, relationship, quality, and the average cost—were determined as 0.45, 0.04, 0.10, 0.31, and 0.10, respectively. Obviously, the

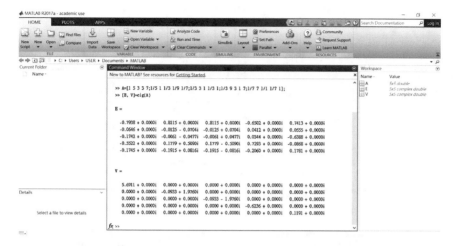

Fig. 4.7   Applying MATLAB R2017a to derive the eigenvalue and eigenvector of **A**

printing speed and quality were two of the most important criteria for the UM system administrator.

A customer placed an order of six pieces of an action figure (see Fig. 4.8) online. After receiving this order, the UM system administrator negotiated with each 3D printing facility in the service region. There were nine 3D printing facilities available for this order. The performances of these 3D printing facilities in optimizing various

Fig. 4.8   The printed action figure [13]

Table 4.2 The performances of 3D printing facilities

3D printing facility (k)	p_{k1}	p_{k2}	p_{k3}	p_{k4}	p_{k5}
1	1.0	1.0	4.0	2.5	2.5
2	2.5	2.5	2.5	4.0	2.5
3	4.0	1.0	2.5	2.5	2.5
4	4.0	4.0	4.0	2.5	2.5
5	5.0	1.0	2.5	4.0	1.0
6	4.0	1.0	2.5	2.5	2.5
7	2.5	1.0	5.0	4.0	1.0
8	4.0	2.5	4.0	4.0	2.5
9	5.0	2.5	2.5	2.5	4.0

criteria were evaluated according to the rules in Table 4.1 and are summarized in Table 4.2.

The normalized performances of the 3D printing facilities are obtained as summarized in Table 4.3. The weights of criteria obtained using AHP were multiplied to the normalized performances to derive the weighted scores. The results are presented in Table 4.4.

The two reference points, the ideal point Λ^+ and the anti-ideal point Λ^-, were established, respectively, as

$$\Lambda_i^+ = [0.451 \ 0.037 \ 0.099 \ 0.314 \ 0.099]$$
$$\Lambda_i^- = [0.090 \ 0.009 \ 0.050 \ 0.196 \ 0.025]$$

The distances from each 3D printing facility to the two reference points were measured, as summarized in Table 4.5.

Based on the two distances, the closeness of each 3D printing facility was calculated. The results are shown in Table 4.6. Since only six pieces were to be printed,

Table 4.3 The normalized performances of 3D printing facilities

3D printing facility (k)	ρ_{k1}	ρ_{k2}	ρ_{k3}	ρ_{k4}	ρ_{k5}
1	0.200	0.250	0.800	0.625	0.625
2	0.500	0.625	0.500	1.000	0.625
3	0.800	0.250	0.500	0.625	0.625
4	0.800	1.000	0.800	0.625	0.625
5	1.000	0.250	0.500	1.000	0.250
6	0.800	0.250	0.500	0.625	0.625
7	0.500	0.250	1.000	1.000	0.250
8	0.800	0.625	0.800	1.000	0.625
9	1.000	0.625	0.500	0.625	1.000

Table 4.4 The weighted scores

3D printing facility (k)	s_{k1}	s_{k2}	s_{k3}	s_{k4}	s_{k5}
1	0.090	0.009	0.079	0.196	0.062
2	0.226	0.023	0.050	0.314	0.062
3	0.361	0.009	0.050	0.196	0.062
4	0.361	0.037	0.079	0.196	0.062
5	0.451	0.009	0.050	0.314	0.025
6	0.361	0.009	0.050	0.196	0.062
7	0.226	0.009	0.099	0.314	0.025
8	0.361	0.023	0.079	0.314	0.062
9	0.451	0.023	0.050	0.196	0.099

Table 4.5 The distances from each 3D printing facility to the two reference points

3D printing facility (k)	d_k^+	d_k^-
1	0.383	0.048
2	0.234	0.184
3	0.163	0.273
4	0.154	0.276
5	0.094	0.380
6	0.163	0.273
7	0.239	0.186
8	0.101	0.299
9	0.128	0.369

Table 4.6 The closeness of each 3D printing facility

3D printing facility (k)	C_k
1	0.111
2	0.439
3	0.626
4	0.642
5	0.802
6	0.626
7	0.438
8	0.749
9	0.742

the top six performing 3D printing facilities were chosen as #5, #8, #9, #4, #3, and #6. However, this result did not mean that all the six 3D printing facilities printed the required pieces, but the six 3D printing facilities were considered in distributing the required pieces, according to the production and transportation planning results.

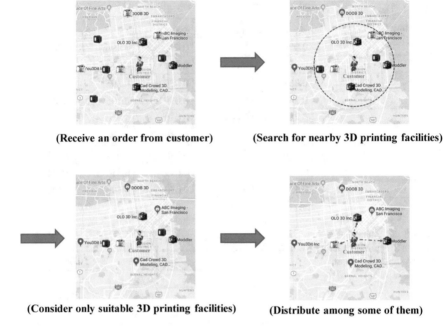

(Receive an order from customer)   **(Search for nearby 3D printing facilities)**

(Consider only suitable 3D printing facilities)   **(Distribute among some of them)**

Fig. 4.9 The process of selecting 3D printing facilities

Therefore, it was possible that some of the six 3D printing facilities did not print any piece.

There are actually four steps in selecting 3D printing facilities, as illustrated in Fig. 4.9.

References

1. F. Tao, Y. Cheng, L.D. Xu, L. Zhang, B.H. Li, CCIoT-CMfg: cloud computing and internet of things-based cloud manufacturing service system. IEEE Trans. Industr. Inf. **10**(2), 1435–1442 (2014)
2. T. Chen, Y.C. Lin, Feasibility evaluation and optimization of a smart manufacturing system based on 3D printing. Int. J. Intell. Syst. **32**, 394–413 (2017)
3. H.C. Wu, T.C.T. Chen, Quality control issues in 3D-printing manufacturing: a review. Rapid Prototyp. J. **24**(3), 607–614 (2018)
4. T. Chen, H.R. Tsai, Ubiquitous manufacturing: current practices, challenges, and opportunities. Robot. Comput.-Integr. Manuf. **45**, 126–132 (2017)
5. T. Chen, Y.C. Wang, An evolving fuzzy planning mechanism for a ubiquitous manufacturing system. Int. J. Adv. Manuf. Technol. (2020) (in press)
6. T.L. Saaty, Decision making with the analytic hierarchy process. Int. J. Serv. Sci. **1**(1), 83–98 (2008)
7. M.C. Lin, C.C. Wang, M.S. Chen, C.A. Chang, Using AHP and TOPSIS approaches in customer-driven product design process. Comput. Ind. **59**(1), 17–31 (2008)

8. Y.J. Lai, T.Y. Liu, C.L. Hwang, Topsis for MODM. Eur. J. Oper. Res. **76**(3), 486–500 (1994)
9. J.W. Wang, C.H. Cheng, K.C. Huang, Fuzzy hierarchical TOPSIS for supplier selection. Appl. Soft Comput. **9**, 377–386 (2009)
10. Y.C. Lin, Y.C. Wang, T.C.T. Chen, H.F. Lin, Evaluating the suitability of a smart technology application for fall detection using a fuzzy collaborative intelligence approach. Mathematics **7**(11), 1097 (2019)
11. G. Büyüközkan, G. Çifçi, A novel hybrid MCDM approach based on fuzzy DEMATEL, fuzzy ANP and fuzzy TOPSIS to evaluate green suppliers. Expert Syst. Appl. **39**(3), 3000–3011 (2012)
12. F.R.L. Junior, L. Osiro, L.C.R. Carpinetti, A comparison between Fuzzy AHP and Fuzzy TOPSIS methods to supplier selection. Appl. Soft Comput. **21**, 194–209 (2014)
13. Y.-C. Lin, T. Chen, A ubiquitous manufacturing network system. Robot. Comput.-Integr. Manuf. **45**, 157–167 (2017)

Chapter 5
Production and Transportation Planning for a Ubiquitous Manufacturing System Based on Three-Dimensional Printing

5.1 A Ubiquitous Manufacturing System Based on Three-Dimensional Printing

Advances in computer and communication technologies have contributed to the rapid development of interactive systems for modeling, simulating, and even prototyping three-dimensional (3D) objects [1]. Traditionally, 3D printing is applied to prototyping. However, some recent attempts have shown that 3D printing can support mass customization and mass production [2–4]. In addition, a 3D printer is easy and cheap to acquire, and the 3D model of a product can be transmitted through the Internet. For these reasons, 3D printing can contribute to ubiquitous manufacturing (UM), so that a product can be manufactured anywhere and at any time by using 3D printing [4]. As a result, a UM system based on 3D printing can be established [5, 6]. Such a UM system is distributively managed based on a service-oriented architecture (SOA) consisting of a collection of loosely coupled 3D printing facilities that communicate with each other through standard interfaces using standard message-exchanging protocols [7] or centrally managed with the intervention of a UM service provider (i.e., the system administrator). Further, business-to-customer (B2C) e-commerce can be incorporated into such a system, in which a customer uses a smartphone, personal computer (PC), or personal digital assistant (PDA) to place an order of 3D objects [8], as illustrated in Fig. 5.1.

This chapter is dedicated to the production planning of a UM system. Dubey et al. [9] analyzed the existing literature on UM using techniques including interpretive structural modeling (ISM) and cross-impact matrix multiplication (MICMAC). Then, they established a three-level framework showing that UM culture, training, secondary technology, pervasive technology, product quality, manufacturing flexibility, and teamwork were critical to the success of a UM system.

When a job is moved from a machine to another, or from a factory to another, it is subject to delays. Possible reasons for the delays include inefficient transportation and the destination not informed of the arrival of the job immediately. To address this,

T.-C. Chen, *3D Printing and Ubiquitous Manufacturing*,
SpringerBriefs in Applied Sciences and Technology,
https://doi.org/10.1007/978-3-030-49150-5_5

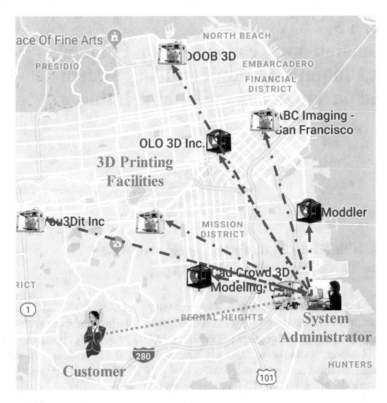

Fig. 5.1 A UM system based on 3D printing

radio frequency identification (RFID) and other types of sensors can be attached to manufacturing resources to monitor and collect real-time production conditions and information, based on which better production planning and scheduling can be made [10–13]. However, existing UM applications were confined to highly automatic factories or functionalities (such as machining, robots, 3D printers, equipment monitoring, and production simulation) that respond to the detected conditions instantaneously. For example, Zhong et al. [12] formulated two nonlinear programming (NLP) models for planning the production and scheduling of a UM system based on the application of RFIDs. The two NLP models aimed to minimize the mean lateness and the total lateness, respectively. The production plan became constraints on the schedule. Heuristics were proposed to help solve the two NLP problems. Chen and Lin [14] established a UM system for distributing production simulation tasks among several cloud-based simulation services. Wang et al. [15] integrated various types of manufacturing facilities with function blocks. Their methodology was successfully applied to machining and robotics applications. Chen and Lin [16] constructed a UM system based on the application of 3D printing (or additive manufacturing). In their UM system, the pieces of an order were spread across multiple 3D printing facilities. The printed pieces were then picked up by a transportation service provider that visited all 3D printing facilities sequentially. Li et al. [17] formulated an NLP model that distributed jobs among 3D printers, so as to minimize total processing costs including material melting costs (depending on the volumes of parts), powder layering costs (depending on the maximum height of parts), and setup costs. Changeovers were not considered because all 3D printers used the same materials. However, efficiency may be more important than cost-effectiveness to a 3D printing-based UM system. Chen and Lin formulated an NLP model to minimize the cycle time for delivering an order, and proposed a heuristic to help solve the NLP problem. Wang et al. [18] established a fault-diagnosis and early-warning mechanism for a high-end assembly system under a UM environment. Compared to a single factory, a UM system that moves jobs across multiple distributed factories is subject to more uncertainty. For example, the transportation time between any two factories depends on the traffic condition. The operations in each factory involve inconsistent human intervention. Such uncertainty can be considered using fuzzy logic [19] or stochastic networks [20].

A UM system composed of distributed 3D printing facilities is the focus of this chapter. In the view of Mai et al. [21], the basic process of such a system is composed of the following steps:

Step 1. 3D printing service matching and optimal selection.
Step 2. Service request.
Step 3. Order confirmation.
Step 4. Business management.
Step 5. Transaction.
Step 6. Service execution.
Step 7. Production management.
Step 8. Logistics.
Step 9. User evaluation.

Chen and Wang [22] established the operational procedure of a UM system based on 3D printing, illustrated in Fig. 5.2, as follows [22]:

(1) A customer places an order of products that can be 3D printed with a smartphone.
(2) Then, the content of the order as well as the customer's location is transmitted to the system server. In theory, the system server can be placed anywhere, or even outside the UM system service region.
(3) The availability of each 3D printing facility is sensed and transmitted to the system server: The possible statuses of a 3D printer are summarized in Table 5.1. A 3D printer is available only when its status is "ready" [22]. If a UM system is distributively managed, then a broker mechanism is required to fulfill this task [8].
(4) The system server searches the system database for 3D printing facilities that are not only available but also close to the customer.
(5) The required pieces are distributed among 3D printing facilities.
(6) Each 3D printing facility prints the required pieces.

Fig. 5.2 The operational
procedure of a UM system
based on 3D printing

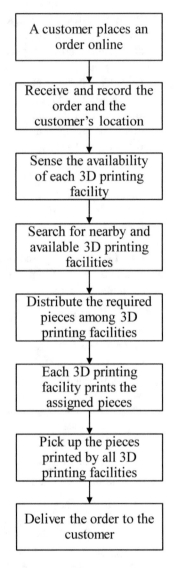

Table 5.1 Possible statuses
of a 3D printer

Status No.	Status
1	The 3D printer has not been initialized
2	The 3D printer is ready
3	The computer is sending/spooling a job to the 3D printer
4	The extruder is heating
5	3D printing is in progress
6	3D printing has stopped

(7) A freight truck visits each 3D printing facility to pick up the printed pieces and deliver them to the customer.

Two challenges faced by the system administrator of a UM system based on 3D printing are balancing the workloads of 3D printing facilities and identifying the shortest delivery path [22]. By contrast, in the UM system considered by Argoneto and Renna [23], it was assumed that factories could exchange their capacity freely to fulfill an order. However, exchanging capacity between factories that were far from each other should be avoided.

5.2 Balancing the Workloads on Three-Dimensional Printing Facilities

Balancing the workloads on 3D printing facilities helps avoid the starvation or congestion of any 3D printing facility. To this end, the number of pieces that are assigned to each 3D printing facility is determined as follows:

$$\text{Min } Z_1 = \max_i (a_i + n_i p_i) \tag{5.1}$$

subject to

$$\sum_{i=1}^{m} n_i = N \tag{5.2}$$

$$n_i \in Z^+ \cup \{0\}; i = 1 \sim m \tag{5.3}$$

where Z_1 is the cycle time for completing the order; a_i is the available time of the i-th 3D printing facility, $i = 1 \sim m$; n_i is the number of pieces to be printed at the i-th 3D printing facility; p_i is the time required to print a piece in the i-th 3D printing facility. In total, there are N pieces to be printed. In theory, the number of feasible solutions is $(N + 1)^m$, which grows rapidly as m increases. To address this, only 3D printing facilities close to the customer were considered, as illustrated in Fig. 5.3.

The objective function minimizes the maximal completion time (i.e., the makespan), which can be replaced by the following constraint:

$$Z_1 \geq a_i + n_i p_i; i = 1 \sim m \tag{5.4}$$

This model is a mixed-integer linear programming (MILP) problem that is NP-hard. The simplest way to solve this type of problem is the LP relaxation, namely, removing the integer assumption and then solving it as a linear programming (LP) problem. The optimal solution to the LP relaxation is then rounded to the nearest

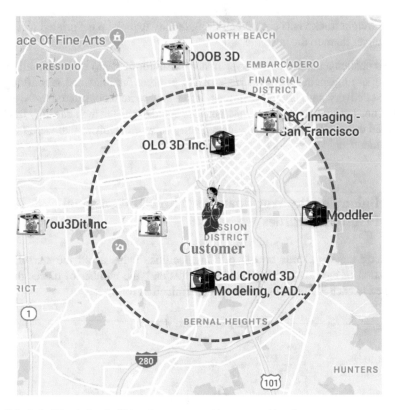

Fig. 5.3 Only 3D printing facilities close to a customer are considered

integer [24]. Alternatively, a branch-and-bound (B&B) algorithm can be designed to
search for the global optimal solution systematically.

5.3   Identifying the Shortest Delivery Path

The shortest path to 3D printing facilities that print at least one piece is determined
by optimizing the following model [22]:

$$\text{Min } Z_2 \tag{5.5}$$

subject to

$$r_i \geq X_{Oi}(t + d_{Oi}); i = 1 \sim m \tag{5.6}$$

$$\sum_{i=1}^{m} X_{Oi} = 1 \tag{5.7}$$

$$r_i \geq X_{ji}(r_j + d_{ji}); i, j = 1 \sim m; j \neq i \tag{5.8}$$

$$r_j \geq X_{ij}(r_i + d_{ij}); i, j = 1 \sim m; j \neq i \tag{5.9}$$

$$X_{ij} + X_{ji} \leq 1; i, j = 1 \sim m; j \neq i; \tag{5.10}$$

$$X_{Oi} + \sum_{j \neq i} X_{ji} = 1; i = 1 \sim m \tag{5.11}$$

$$X_{iO} + \sum_{j \neq i} X_{ij} = 1; i = 1 \sim m \tag{5.12}$$

$$Z_2 \geq X_{iO}(r_i + d_{iO}); i = 1 \sim m \tag{5.13}$$

$$\sum_{i=1}^{m} X_{iO} = 1 \tag{5.14}$$

$$X_{Oi}, X_{iO}, X_{ij} \in \{0, 1\}; i, j = 1 \sim m; j \neq i \tag{5.15}$$

where $r_{(i)}$ is the arrival time at the i-th 3D printing facility; O indicates the start location as well as the destination; t is the current time; d_{Oi} is the distance between O and 3D printing facility i; d_{ij} is the distance between 3D printing facilities i and j, which is set to the length of the shortest path between them. If two pieces are made by the same 3D printing facility, then $i = j$ and $d_{ij} = 0$. If the freight truck goes from O to 3D printing facility j, $X_{Oi} = 1$; otherwise, $X_{Oi} = 0$. Similarly, if the freight truck goes from 3D printing facility i to 3D printing facility j, $X_{ij} = 1$; otherwise, $X_{ij} = 0$. Moreover, obviously,

$$d_{ij} = d_{ji} \tag{5.16}$$

The objective function minimizes the length of the delivery path [25, 26]. Constraints (5.6) and (5.7) request that the freight truck begins from O and moves toward one of the 3D printing facilities. Constraints (5.8)–(5.10) show the two possible directions of moving between 3D printing facilities i and j. The distance between two 3D printing facilities d_{ji} is expressed in terms of the required transportation time and therefore can be added up with the arrival time r_j. Constraints (5.11) and (5.12) request that a 3D printing facility is visited once. Then, the freight truck continues to one of the other 3D printing facilities and finally returns to O as required by constraint (5.13).

This model is a mixed-integer quadratic programming (MIQP) problem that is NP-hard [25]. Relaxing the integer variables leads to a quadratic programming (QP) problem, which is a slightly easier but also NP-hard if the constraints are not convex. Another B&B algorithm can be designed to help find the global optimal solution. In contrast, Chen and Lin [16] applied Google maps to achieve the same purpose.

5.4   An Aggregate Production and Transportation Planning Model

The two previous models can be merged as follows:

$$\text{Min } Z_3 \tag{5.17}$$

subject to

$$r_i \geq X_{Oi}(t + d_{Oi}); i = 1 \sim m \tag{5.18}$$

$$\sum_{i=1}^{m} X_{Oi} = 1 \tag{5.19}$$

$$c_i = a_i + n_i p_i; i = 1 \sim m \tag{5.20}$$

$$\sum_{i=1}^{m} n_i = N \tag{5.21}$$

$$l_i \geq r_i; i = 1 \sim m \tag{5.22}$$

$$l_i \geq c_i; i = 1 \sim m \tag{5.23}$$

$$r_i \geq X_{ji}(l_j + d_{ji}); i, j = 1 \sim m; j \neq i \tag{5.24}$$

$$r_j \geq X_{ij}(l_i + d_{ij}); i, j = 1 \sim m; j \neq i \tag{5.25}$$

$$X_{ij} + X_{ji} \leq 1; i, j = 1 \sim m; j \neq i \tag{5.26}$$

$$X_{Oi} + \sum_{j \neq i} X_{ji} \leq 1; i = 1 \sim m \tag{5.27}$$

$$X_{Oi} + \sum_{j \neq i} X_{ji} \geq 0^+ n_i; i = 1 \sim m \tag{5.28}$$

$$X_{iO} + \sum_{j \neq i} X_{ij} = X_{Oi} + \sum_{j \neq i} X_{ji} \tag{5.29}$$

$$Z_3 = \sum_{i=1}^{m} X_{iO}(l_i + d_{iO}); i = 1 \sim m \tag{5.30}$$

$$\sum_{i=1}^{m} X_{iO} = 1 \tag{5.31}$$

$$X_{Oi}, X_{iO}, X_{ij} \in \{0, 1\}; i, j = 1 \sim m; j \neq i \tag{5.32}$$

$$n_i \in Z^+ \cup \{0\} \tag{5.33}$$

where c_i is the completion time at the i-th 3D printing facility; l_i is the time when the freight truck leaves the i-th 3D printing facility. Constraint (5.28) forces the freight truck to visit a 3D printing facility that prints at least one piece; 0^+ is a very small positive value. In Eq. (5.30), the distance between a 3D printing facility and the start location d_{iO} is expressed in terms of the required transportation time, and therefore can be added to the leaving time l_i. This model is also an MIQP problem.

A B&B algorithm can be applied to help solve the MIQP problem. The following properties are useful for establishing bounds in the B&B algorithm [22].

Property 1
Both Z_1^* and Z_2^* serve as lower bounds on Z_3^*.

Property 2
The optimal objective function value of the LP relaxation of the workload balancing model, indicated by Z_1^*-rlx, serves as a lower bound on Z_3^*.

Property 3
The optimal objective function value of the QP relaxation of the transportation planning model, indicated by Z_2^*-rlx, serves as a lower bound on Z_3^*.

Based on these properties, the B&B algorithm shown in Fig. 5.4 can be applied to search for the global optimal solution to the optimization problem.

In contrast, Chen and Lin [16] proposed a heuristic to find a locally optimal solution. Wu et al. [20] constructed a stochastic Petri net (SPN) model to help solve the problem. Wu et al.'s method split an order among all 3D printing facilities, rather than just some of them. Such a treatment is more suitable for cases in which all 3D printing facilities are equipped with the same 3D printers.

If the 3D objects printed by different 3D printing facilities are the parts of a single product and need to be assembled, then there is an issue of bottleneck [20]. In such

Initialize: Incumbent := ∞; LB(P_0) := max(Z_1^* -rlx(P_0), Z_2^* -rlx(P_0)); Live := {(P_0 , LB(P_0))}
Repeat until Live = ∅
    Select the node P from Live to be processed; Live := Live \ {P};
    Branch on P generating P_1, ..., P_k ;
    For $1 \leq l \leq k$ do
        Bound P_l : LB(P_l) := max(Z_1^* -rlx(P_l), Z_2^* -rlx(P_l));
        If LB(P_l) = $f(X)$ for a feasible solution X and $f(X)$ < Incumbent then
            Incumbent := $f(X)$; Solution := X;
            go to EndBound;
        If LB(P_l) ≥ Incumbent then fathom P_l
            else Live := Live ∪ {(P_l , LB(P_l))}
    EndBound;
OptimalSolution := Solution; OptimumValue := Incumbent

Fig. 5.4 The B&B algorithm [22]

a situation, the progresses at all 3D printing facilities have to be coordinated, so that the printed parts do not need to wait for parts in process. This problem is even more complicated after considering the transportation time between each of the two 3D printing facilities. Nevertheless, if the printed parts are assembled at the destination, the proposed aggregate production and planning model still gives the optimal solution that minimizes the delivery time to the customer.

5.5 An Example

An example is used to illustrate the applicability of the proposed methodology. In this example, the service region has an area of approximately 26 km^2. A customer placed an order of three pieces of a 3D object through a smartphone. The customer's location was detected using the global positioning system (GPS) module on his smartphone. Then, the customer's order and detected location were transmitted to the system server. After receiving this information, the system server searched the system database for nearby 3D printing facilities to print the order collaboratively.

There were six 3D printing facilities in the proximity of the customer. The details about the facilities are presented in Table 5.2. The distance between any two 3D printing facilities, in terms of the required travel time, was estimated using Google Maps. Such an approach is also more precise in practice. The results formed a distance matrix, which is revealed in Table 5.3. Distance matrixes are a critical part of the production and transportation planning of a 3D printing-based UM system [19, 20]. Constructing distance matrixes in advance by considering the average traffic conditions can facilitate the subsequent decision-making process. At the beginning of the experiment, all six 3D printing facilities were available.

The time required for printing one piece of the 3D object at each 3D printing facility is shown in Table 5.4. The distance from the customer to each 3D printing facility was also estimated using Google Maps. The results are summarized in Table 5.5.

Table 5.2 Details about the 3D printing facilities

3D printing facility	Products or services provided
A	• 3D figure design • 3D printing service
B	• 3D printer • 3D scanner • 3D printing service
C	• 3D printing service
D	• 3D modeling • 3D printing service
E	• 3D printer • 3D scanner • 3D printing service • 3D reversed engineering
F	• 3D modeling • 3D printing service

Table 5.3 The distance matrix (unit: min)

3D printing facility	A	B	C	D	E	F
A	0	13	26	18	13	16
B	13	0	29	18	16	19
C	26	29	0	34	20	11
D	18	18	34	0	23	20
E	13	16	20	23	0	7
F	16	19	11	20	7	0

Table 5.4 The time required for printing one piece (unit: min)

3D printing facility	Unit printing time
A	88
B	88
C	71
D	88
E	71
F	71

The MILP model for distributing the required pieces among the 3D printing facilities was coded using Lingo, as shown in Fig. 5.5. The optimal solution was $\{n_i\} = \{0, 0, 1, 0, 1, 1\}$, and the optimal objective function value was $Z_1^* = 78$. In other words, the required pieces could be finished in 78 min through the collaboration of 3D printing facilities C, E, and F.

Table 5.5 The distance from the customer to each 3D printing facility (unit: min)

3D Printing Facility	Distance
A	32
B	27
C	47
D	20
E	35
F	38

```
min=Z1;
Z1>=0+n1*104;
...
Z1>=0+n6*78;
n1+n2+n3+n4+n5+n6=3;
@gin(n1); ...; @gin(n6);
```

Fig. 5.5 The MILP model

```
min=Z2;
r3>=XO3*0+XO3*47;
...
r6>=XO6*0+XO6*38;
XO3+XO5+XO6=1;
r3>=X53*r5+X53*20;
r5>=X35*r3+X35*20;
X35+X53<=1;
...
r5>=X65*r6+X65*7;
r6>=X56*r5+X56*7;
X56+X65<=1;
XO3+X53+X63=1;
...
XO6+X56+X36=1;
X3O+X35+X36=1;
...
X6O+X63+X65=1;
Z2>=X3O*r3+X3O*47;
...
Z2>=X6O*r6+X6O*38;
X3O+X5O+X6O=1;
@bin(X35); ...; @bin(X65);
@bin(X3O); ...; @bin(X6O);
@bin(XO3); ...; @bin(XO6);
```

Fig. 5.6 The MIQP model

Subsequently, the MIQP model for finding the shortest delivery path through the three 3D printing facilities was also coded using Lingo, as shown in Fig. 5.6. The shortest delivery path was O (start location) $\rightarrow$ C $\rightarrow$ F $\rightarrow$ E $\rightarrow$ O, and the optimal objective function value was $Z_2^* = 100$. The results of the two models were then combined to form the production and transportation plan (see Table 5.6).

Notably, the cycle time required for fulfilling the order was 2017/7/9 11:35 − 2017/7/9 09:24 = 131 (min), which was greater than Z_1^* or Z_2^* but much less than $Z_1^* + Z_2^*$, showing the benefits of shortening the cycle time through the collaboration of the three 3D printing facilities.

To further improve the planning performance, the aggregate production and transportation planning model was formulated and optimized with the aid of the B&B algorithm, as shown in Fig. 5.7. The results are as follows:

(1) The optimization objective function value was $Z_3^* = 125$ when $\{n_i\} = \{0, 0, 0, 1, 1, 1\}$ and the delivery path was O (start location) $\rightarrow$ E $\rightarrow$ F $\rightarrow$ D $\rightarrow$ O.
(2) After enumerating all potentially feasible solutions to make a comparison, the solution obtained using the proposed B&B algorithm was determined to be the global optimal solution.
(3) The planning performance was substantially enhanced from that when the two problems were addressed separately. The production and transportation plan is shown in Table 5.7. The total cycle time was shortened from 131 min to 125 min.

If the three items are different parts of a product that needed to be assembled at the destination, the production and planning plan was also optimal by minimizing the cycle time.

Table 5.6 The production and transportation plan

Time	Event
2017/7/9 09:24	• The order was received • 3D printing facilities C, E, and F started to print the required items • The freight truck left the start location and began traveling to 3D printing facility C
2017/7/9 10:11	• The freight truck arrived at 3D printing facility C
2017/7/9 10:42	• 3D printing facility C printed 1 item • The freight truck picked up the item printed by 3D printing facility C and moved to 3D printing facility F • 3D printing facility E printed 1 item • 3D printing facility F printed 1 item
2017/7/9 10:53	• The freight truck arrived at 3D printing facility F, picked up the printed item, and moved to 3D printing facility E
2017/7/9 11:00	• The freight truck arrived at 3D printing facility E, picked up the printed item, and began traveling back to the start location
2017/7/9 11:35	• The freight truck arrived at the start location

```
min=Z3;
r1>=XO1*0+XO1*32;
...
r3>=XO3*0+XO3*47;
XO1+XO2+XO3+XO4+XO5+XO6=1;
n1+n2+n3+n4+n5+n6=3;
l1>=r1;
l1>=0+n1*104;
...
l6>=r6;
l6>=0+n6*78;
r1>=X21*l2+X21*l3;
r2>=X12*l1+X12*l3;
X12+X21<=1;
...
r5>=X65*l6+X65*7;
r6>=X56*l5+X56*7;
X56+X65<=1;
XO1+X21+X31+X41+X51+X61<=1;
...
XO6+X16+X26+X36+X46+X56<=1;
XO1+X21+X31+X41+X51+X61>=0.01*n1;
...
XO6+X16+X26+X36+X46+X56>=0.01*n6;
X1O+X12+X13+X14+X15+X16=XO1+X21+X31+X41+X51+X61;
...
X6O+X61+X62+X63+X64+X65=XO6+X16+X26+X36+X46+X56;
Z3=X1O*l1+X1O*32+X2O*l2+X2O*27+X3O*l3+X3O*47+X4O*l4+X4O*20+X5O*l5+X5O*35
+X6O*l6+X6O*38;
X1O+X2O+X3O+X4O+X5O+X6O=1;
@bin(XO1); ...; @bin(XO6);
@bin(X1O); ...; @bin(X6O);
@bin(X12); ...; @bin(X65);
@gin(n1); ...; @gin(n6);
```

Fig. 5.7 The aggregate production and transportation model

5.6 A Production Planning Model Considering Uncertainty

The production planning of a UM system based on 3D printing is subject to much uncertainty. To address this issue, uncertain parameters and/or variables can be modeled with fuzzy values, and the following fuzzy mixed-integer linear programming (FMILP) model can be optimized to distribute an order among multiple 3D printing facilities:

$$\text{Min } \tilde{Z}_4 = \max_i (\tilde{a}_i (+) n_i \tilde{p}_i) \qquad (5.34)$$

$$\sum_{i=1}^{m} n_i = N \qquad (5.35)$$

$$n_i \in Z^+ \cup \{0\}; i = 1 \sim m \qquad (5.36)$$

Table 5.7 The production and transportation plan generated by the aggregate model

Time	Event
2017/7/9 09:24	• The order was received • 3D printing facilities D, E, and F started to print the required items • The freight truck left the start location and began traveling to 3D printing facility E
2017/7/9 09:59	• The freight truck arrived at 3D printing facility E
2017/7/9 10:42	• 3D printing facility E printed 1 item • 3D printing facility F printed 1 item • The freight truck picked up the item printed by 3D printing facility E and moved to 3D printing facility F
2017/7/9 10:49	• The freight truck arrived at 3D printing facility F, picked up the printed item, and moved to 3D printing facility D
2017/7/9 11:08	• 3D printing facility D printed 1 item
2017/7/9 11:09	• The freight truck arrived at 3D printing facility D, picked up the printed item, and began traveling back to the start location
2017/7/9 11:29	• The freight truck arrived at the start location

where $\tilde{a}_i$ is the available time of the i-th 3D printing facility; $i = 1 \sim m$. $\tilde{a}_i$ is uncertain because it is difficult to estimate the completion time of a piece because of human intervention. (+) denotes fuzzy multiplication. $\tilde{p}_i$ is the time required to print a piece in the i-th 3D printing facility. $\tilde{p}_i$ is uncertain because of human-assisted operations. $\tilde{a}_i(+)n_i\tilde{p}_i$ is the (fuzzy) completion time at the i-th 3D printing facility that prints n_i pieces. A total of N pieces are to be printed, as depicted by Eq. (5.35).

The objective function $\tilde{Z}_4 = (Z_{41}, Z_{42}, Z_{43})$ is to minimize the fuzzy maximal completion time (or the fuzzy makespan). In the literature, various methods for handling such a fuzzy objective function have been proposed. For example, Chen [26] replaced a fuzzy objective function with its center of gravity (COG), which is simple yet effective in most applications:

$$\text{Min} \frac{Z_{41} + Z_{42} + Z_{43}}{3} \tag{5.37}$$

However, ties form easily in this way. Hsu and Wang [27] split a fuzzy objective function into three crisp objective functions:

$$\text{Min } Z_{42} \tag{5.38}$$

$$\text{Max } Z_{42} - Z_{41} \tag{5.39}$$

$$\text{Min } Z_{43} - Z_{42} \tag{5.40}$$

Subsequently, the desirable region for each objective function is specified to evaluate the satisfaction level, as illustrated in Fig. 5.8:

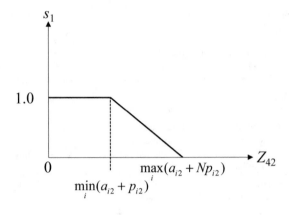

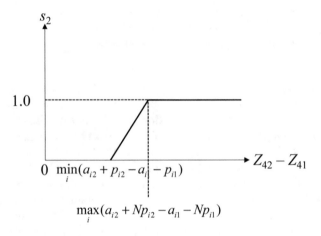

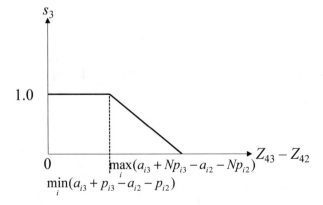

Fig. 5.8 Desirable regions [19]

$$s_1 = \max(\min(\frac{\max_i(a_{i2} + Np_{i2}) - Z_{42}}{\max_i(a_{i2} + Np_{i2}) - \min_i(a_{i2} + p_{i2})}, \; 1), \; 0) \tag{5.41}$$

$$s_2 = \max(\min(\frac{Z_{42} - Z_{41} - \min_i(a_{i2} + p_{i2} - a_{i1} - p_{i1})}{\max_i(a_{i2} + Np_{i2} - a_{i1} - Np_{i1}) - \min_i(a_{i2} + p_{i2} - a_{i1} - p_{i1})}, \; 1), \; 0) \tag{5.42}$$

$$s_3 = \max(\min(\frac{\max_i(a_{i3} + Np_{i3} - a_{i2} - Np_{i2}) - (Z_{43} - Z_{42})}{\max_i(a_{i3} + Np_{i3} - a_{i2} - Np_{i2}) - \min_i(a_{i3} + p_{i3} - a_{i2} - p_{i2})}, \; 1), \; 0) \tag{5.43}$$

Subsequently, the three satisfaction levels are aggregated as

$$\text{Max } s_{\min} = \min(s_1, \; s_2, \; s_3) \tag{5.44}$$

Finally, the following MILP model is optimized instead:

$$\text{Max } s_{\min} \tag{5.45}$$

subject to

$$s_{\min} \leq s_1 \tag{5.46}$$

$$s_{\min} \leq s_2 \tag{5.47}$$

$$s_{\min} \leq s_3 \tag{5.48}$$

$$s_1 \leq \frac{\max_i(a_{i2} + Np_{i2}) - Z_{42}}{\max_i(a_{i2} + Np_{i2}) - \min_i(a_{i2} + p_{i2})} \tag{5.49}$$

$$0 \leq s_1 \leq 1 \tag{5.50}$$

$$s_2 \leq \frac{Z_{42} - Z_{41} - \min_i(a_{i2} + p_{i2} - a_{i1} - p_{i1})}{\max_i(a_{i2} + Np_{i2} - a_{i1} - Np_{i1}) - \min_i(a_{i2} + p_{i2} - a_{i1} - p_{i1})} \tag{5.51}$$

$$0 \leq s_2 \leq 1 \tag{5.52}$$

$$s_3 \leq \frac{\max_i(a_{i3} + Np_{i3} - a_{i2} - Np_{i2}) - (Z_{43} - Z_{42})}{\max_i(a_{i3} + Np_{i3} - a_{i2} - Np_{i2}) - \min_i(a_{i3} + p_{i3} - a_{i2} - p_{i2})} \tag{5.53}$$

$$0 \leq s_3 \leq 1 \tag{5.54}$$

$$Z_{41} \geq a_{i1} + n_i\, p_{i1}, i = 1 \sim m \tag{5.55}$$

$$Z_{42} \geq a_{i2} + n_i\, p_{i2}, i = 1 \sim m \tag{5.56}$$

$$Z_{43} \geq a_{i3} + n_i\, p_{i3}, i = 1 \sim m \tag{5.57}$$

$$\sum_{i=1}^{m} n_i = N \tag{5.58}$$

$$0 \leq Z_{41} \leq Z_{42} \leq Z_{43} \tag{5.59}$$

$$n_i \in Z^{+} \cup \{0\}; i = 1 \sim m \tag{5.60}$$

References

1. M.L. Xu, N.B. Gu, W.W. Xu, M.Y. Li, J.X. Xue, B. Zhou, Mechanical assembly packing problem using joint constraints. J. Comput. Sci. Technol. **32**(6), 1162–1171 (2017)
2. C. Schubert, M.C. Van Langeveld, L.A. Donoso, Innovations in 3D printing: a 3D overview from optics to organs. Br. J. Ophthalmol. **98**(2), 159–161 (2014)
3. T. Chen, Y.C. Lin, Feasibility evaluation and optimization of a smart manufacturing system based on 3D printing. Int. J. Intell. Syst. **32**, 394–413 (2017)
4. T. Chen, H.R. Tsai, Ubiquitous manufacturing: current practices, challenges, and opportunities. Robot. Comput. Integr. Manuf. **45**, 126–132 (2017)
5. I. Ryoo, K. Sun, J. Lee, S. Kim, A 3-dimensional group management MAC scheme for mobile IoT devices in wireless sensor networks. J. Ambient Intell. Humaniz. Comput. **9**, 1223–1234 (2018)
6. K. Ko, T. Kim, H. Kim, Management platform of threats information in IoT environment. J. Ambient Intell. Humaniz. Comput. **9**, 1167–1176 (2018)
7. X.V. Wang, L. Wang, A. Mohammed, M. Givehchi, Ubiquitous manufacturing system based on cloud: a robotics application. Robot. Comput. Integr. Manuf. **45**, 116–125 (2017)
8. B. Yoo, H. Ko, S. Chun, Prosumption perspectives on additive manufacturing: reconfiguration of consumer products with 3D printing. Rapid Prototyping J **22**(4), 691–705 (2016)
9. R. Dubey, A. Gunasekaran, A. Chakrabarty, Ubiquitous manufacturing: overview, framework and further research directions. Int. J. Comput. Integr. Manuf. **30**(4–5), 381–394 (2017)
10. J. Fang, T. Qu, Z. Li, G. Xu, G.Q. Huang, Agent-based gateway operating system for RFID-enabled ubiquitous manufacturing enterprise. Robot. Comput. Integr. Manuf. **29**(4), 222–231 (2013)
11. H. Luo, J. Fang, G.Q. Huang, Real-time scheduling for hybrid flowshop in ubiquitous manufacturing environment. Comput. Ind. Eng. **84**, 12–23 (2015)
12. R.Y. Zhong, G.Q. Huang, S. Lan, Q.Y. Dai, T. Zhang, C. Xu, A two-level advanced production planning and scheduling model for RFID-enabled ubiquitous manufacturing. Adv. Eng. Inform. **29**(4), 799–812 (2015)
13. L. Hu, T. Peng, C. Peng, R. Tang, Energy consumption monitoring for the order fulfilment in a ubiquitous manufacturing environment. Int. J. Adv. Manuf. Technol. **89**(9–12), 3087–3100 (2017)

14. T. Chen, C.-W. Lin, Estimating the simulation workload for factory simulation as a cloud service. J. Intell. Manuf. **28**, 1139–1157 (2017)
15. J. Wang, J. Xie, R. Zhao, L. Zhang, L. Duan, Multisensory fusion based virtual tool wear sensing for ubiquitous manufacturing. Robot. Comput. Integr. Manuf. **45**, 47–58 (2017)
16. T.C.T. Chen, Y.C. Lin, A three-dimensional-printing-based agile and ubiquitous additive manufacturing system. Robot. Comput. Integr. Manuf. **55**, 88–95 (2019)
17. Q. Li, I. Kucukkoc, D.Z. Zhang, Production planning in additive manufacturing and 3D printing. Comput. Oper. Res. **83**, 157–172 (2017)
18. M. Wang, Z. Zhang, K. Li, Z. Zhang, Y. Sheng, S. Liu, Research on key technologies of fault diagnosis and early warning for high-end equipment based on intelligent manufacturing and Internet of Things. Int. J. Adv. Manuf. Technol. 1–10
19. T.C.T. Chen, Fuzzy approach for production planning by using a three-dimensional printing-based ubiquitous manufacturing system. AI EDAM **33**(4), 458–468 (2019)
20. D. Wu, D.W. Rosen, D. Schaefer, Scalability planning for cloud-based manufacturing systems. J. Manuf. Sci. Eng. **137**(4), 041007 (2015)
21. J. Mai, L. Zhang, F. Tao, L. Ren, Customized production based on distributed 3D printing services in cloud manufacturing. Int. J. Adv. Manuf. Technol. **84**(1–4), 71–83 (2016)
22. T. Chen, Y.C. Wang, An advanced IoT system for assisting ubiquitous manufacturing with 3D printing. Int. J. Adv. Manuf. Technol. **103**(5–8), 1721–1733 (2019)
23. P. Argoneto, P. Renna, Supporting capacity sharing in the cloud manufacturing environment based on game theory and fuzzy logic. Enterp. Inf. Syst. **10**(2), 193–210 (2016)
24. J.K. Karlof, *Integer Programming: Theory and Practice* (CRC Press, Boca Raton, FL, 2006)
25. B.V. Cherkassky, A.V. Goldberg, T. Radzik, Shortest paths algorithms: theory and experimental evaluation, in *ACM-SIAM Symposium on Discrete Algorithms* (1994), pp 516–525
26. T. Chen, A fuzzy mid-term single-fab production planning model. J. Intell. Manuf. **14**, 273–285 (2003)
27. H.M. Hsu, W.P. Wang, Possibilistic programming in production planning of assemble-to-order environments. Fuzzy Sets Syst. **119**(1), 59–70 (2001)

Chapter 6
Quality Control in a 3D Printing-Based Ubiquitous Manufacturing System

6.1 Quality and Quality Control

The quality of a product can be assessed along various dimensions: performance, reliability, durability, serviceability, aesthetics, features, perceived quality, and conformance to specifications or standards, as illustrated in Fig. 6.1 [1]. The definitions of these quality dimensions are given in Table 6.1.

To ensure the quality of products made in a factory, quality control (QC) techniques are adopted. According to the American Society for Quality [2], QC includes the observation techniques and activities used to fulfill quality requirements. The most prevalent QC techniques include cause-and-effect diagram, check sheet, control chart, histogram, Pareto chart, scatter diagram, and design of experiment (DOE), which are called the seven basic tools of QC [2]. In a QC cycle, QC techniques are applied to the stages of a product life cycle: product design, process planning, IQC, IPQC, and OQC, as shown in Fig. 6.2.

Quality management is to identify and follow quality requirements, audit the results of quality control measurements, use quality measurements to control quality, and recommend changes if necessary [3]. Quality management comprises four main activities: quality planning, quality assurance, quality control, and quality improvement [4].

6.2 Quality of a Three-Dimensional Printed Object

3D printing is used to build a 3D object from a 3D model layer by layer with resin, metal, or other materials [5]. 3D printing has been used to fabricate prototypes, mockups, replacement parts, dental crowns, artificial limbs, and even bridges. Many successful applications have shown that 3D printing is a convenient tool for producing complex internal and external porous structures [6]. 3D printing is also cheaper

© The Author(s), under exclusive license to Springer Nature Switzerland AG 2020
T.-C. Chen, *3D Printing and Ubiquitous Manufacturing*,
SpringerBriefs in Applied Sciences and Technology,
https://doi.org/10.1007/978-3-030-49150-5_6

Fig. 6.1 The six dimensions of product quality

Table 6.1 Definitions of quality dimensions

Quality dimension	Definition
Performance	A product's primary operating characteristics
Reliability	The likelihood that a product will not fail within a specific time period
Durability	The length of a product's life
Serviceability	The speed with which a product can be put into service when it breaks down
Aesthetics	The user's personal preference
Features	Additional characteristics that enhance the appeal of a product to the user
Perceived quality	The quality attributed to a product based on indirect measures
Conformance to specifications or standards	The precision with which a product meets the specified standards

and more efficient than other rapid prototyping technologies, such as selective laser sintering (SLS) and stereolithography [7].

The quality of 3D-printed objects is a critical factor in the widespread applications of low-cost 3D printing [8]. So far, there has been considerable progress in the quality of 3D-printed products. For example, product quality has been considered as a major limitation in applying 3D printing to biomedical applications [9]. Nevertheless, a drug made through 3D printing was approved by FDA in 2015 [10], which was not easy because of the high standards for making drugs.

Fig. 6.2 A QC cycle

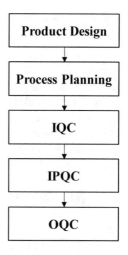

The traditional quality dimensions are not equally emphasized in 3D printing. Aesthetics, conformance to specifications, and performance are three quality dimensions that are more valued [11], as described in subsequent sections.

6.2.1   Quality and QC Standards for 3D Printing

Both .STL and .OBJ are standard file formats for 3D printing, and adhering to them ensures ubiquitous printability for 3D objects [12]. Some standards and guidelines, such as the 3D Printing & Additive Manufacturing Equipment Guideline by UL [13], can be followed while developing 3D printing equipment. Following such standards is believed to yield safe and high-quality equipment.

Some standards and guidelines for the quality of a 3D-printed object and the execution of QC activities within a 3D printing process have recently been proposed by various international organizations. For example, American Society for Testing and Materials International proposed some guidelines for fabricating safe and high-quality components by using powder bed fusion methods involving laser and electron beam sources [14], such as electron beam melting, SLS, selective laser melting, and direct metal laser sintering. Some examples are as follows. ASTM F3091 standard specifies the quality requirements for the mechanical, tolerance, surface finishing, and postprocessing properties of polyamide components fabricated through SLS. ASTM F3049 standard is a version of ASTM F3091 standard that is tailored for the automotive, aerospace, and medical industries. WK46188 standard instructs how to determine the values of the process parameters for powder bed fusion. However, further efforts for standardization in 3D printing are necessary and underway.

6.2.2 Aesthetics

A 3D-printed object usually has a rough surface finish, owing to the "stair casing effect" [15]. In existing 3D printing technologies, fused deposition modeling (FDM) tends to generate a poor surface finish [16]. In addition, the coloration may not be very clear, making it fail to meet aesthetic requirements. The main reason for this result is the resolution of a 3D printer, or the layer thickness it can achieve [17, 18]. To solve this problem, German [19] used a mix of two types of powder such that the finer powder appeared on the upper surface to generate a finer surface finish. In addition, various power densities can be tested, and the one contributing to the best surface finish will be chosen [19], which requires the application of DOE techniques. Choosing an appropriate orientation also helps to improve the surface finish of a 3D object [16, 20]. Further, the type of printhead, the drop-on-demand type or the continuous jet type, and the slicing program logic influence the surface finish of a 3D-printed object as well [17, 21].

After printing, postprocessing treatments are required to polish the surface [22]. For example, Alfieri et al. [20] proposed a laser processing approach for reducing surface roughness that incorporated scanning optics and beam wobbling. Specifically, metal parts were postprocessed through the selective laser melting of stainless steel. Existing postprocessing treatments to enhance the surface finish of a 3D-printed object include machining (turning, milling, computer numeric control (CNC) machining), abrasive machining, chemical machining, laser surface finishing, and abrasive flow machining [23]. A review of these postprocessing techniques refers to [24].

However, postprocessing treatments are not value-added activities and belong to the seven wastes in lean manufacturing [25], as illustrated in Fig. 6.3. Nevertheless, the application of 3D printing can eliminate wastes including transportation, unnecessary motions, overproduction, and excessive inventory. In addition, as noted by Du Preez et al. [26], postprocessing tasks may increase the level of submicrometer-scale particles and/or organic vapors, a portion of which enters the user's breathing zone, which becomes an occupational health issue.

6.2.3 Conformance to Specifications

Conformance to specifications is the most emphasized quality dimension in existing 3D-printing applications. If an object is scanned by a 3D scanner and then printed by a 3D printer, the shape of the 3D-printed object should be as close as possible to the original one. For example, in diagnosis and treatment planning, 3D biomedical models must be very accurate to be usable [7]. Because the 3D model of an object can be easily shared via the Internet, in theory the object can be reproduced very similarly anywhere if the same printer, material, and printing conditions are used. However, in

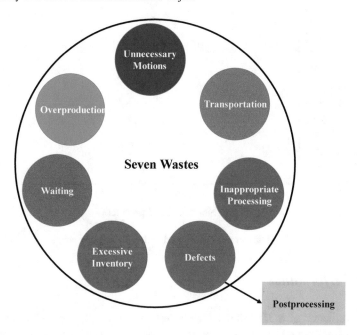

Fig. 6.3 Postprocessing as a waste in lean manufacturing

practice, the reproducibility of an object using 3D printing is still limited, as noted by some studies [27].

A 3D-printed object can be scanned. Then, the scanned model is compared with the original model to measure the deviations [28]. Small deviations are usually required for human organ biofabrication using 3D printing [29]. When the average deviation is smaller than a threshold, it is concluded that the 3D-printed object conforms to the specifications, i.e., a successful printing. Subsequently, another index for the conformance to specifications, yield, can be evaluated. Yield is the number of jobs that have been successfully finished [30]. In a traditional manufacturing process, yield is subject to misoperations. By contrast, 3D printing prevents mistakes by eliminating manual operations, thereby increasing yield [27]. However, products printed with a 3D printer may be considerably different, which is not conducive to the accumulation of knowledge. Consequently, the yield of printing products with a 3D printer may fluctuate. In addition, the yield learning process associated with a 3D printing process may differ from that associated with a traditional, mass production process [30, 31]. Nevertheless, knowledge about the operation of a 3D printer and the control of a 3D-printing process can be accumulated. According to Grieser [32], the yield learning curve associated with a 3D printer (or printing process) may be steep, meaning that such knowledge or experience is easy to accumulate, as illustrated in Fig. 6.4, which means that after a short period of time, a 3D printer (or printing process) will less frequently print poor-quality products. In addition, a user's electronics or mechanical expertise also helps steepen the learning curve [33].

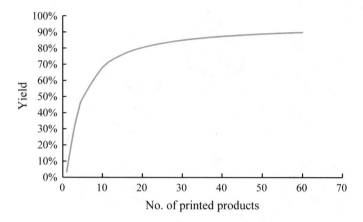

Fig. 6.4 The steep yield learning process associated with a 3D printing process

6.2.4 Performance

The performance of a 3D-printed object depends on its purpose. If a 3D-printed object is only a prototype, it may not have all of the required functions; otherwise, its performance should be comparable to that fabricated using traditional manufacturing technologies. For example, a product traditionally made from molding but now printed with a 3D printer is expected to have features such as high impact resistance and high Young's modulus [34]. A mechanical part, such as a build tray, should have high tensile strength and modulus, even if it is built through 3D printing [35]. Lenses created through 3D printing should achieve high transparency and surface smoothness [22]. Printing the complex vasculature that can supply nutrients to densely populated tissues is one of the biggest roadblocks to generating functional tissues. 3D-printed functional tissues should be able to breathe like healthy tissues in our bodies [36].

Some researchers asserted that a product created through 3D printing is not comparable in functionality to its counterparts made using traditional manufacturing technologies [30]. By contrast, other researchers claimed that 3D printing provides opportunities for further improving the quality of a product. For example, an aircraft can be lightened by 50% by adopting 3D-printed parts instead because of the improved precision in forming parts and the elimination of assembly operations [37].

Usually, the performance of a product can be evaluated from different points of view. However, in bioprinting human organs, Mironov et al. [29] asserted that focusing on a single performance (or functionality) of the printed human organ can prevent failures. Nevertheless, more functionalities and more flexible functionalities are always being pursued.

6.3 QC for 3D Printing

From a manufacturing perspective, 3D printing is a special manufacturing process in which there is no time gap between the research and development (R&D) stage and the mass production stage. In conventional manufacturing processes, a time gap is usually required for building up factory capacity and acquiring raw materials. As a result, traditional QC activities are not equally emphasized in a 3D printing process.

QC aims to manufacture products economically by eliminating defects and waste. To this end, 3D printing is also an effective means, because it reduces the investment in machines, tools, assembly, and materials for diversified products [27]. The combination of QC and 3D achieves a greater synergy. For example, Mironov et al. [29] asserted that an automated QC system effectively improves the quality of 3D-printed human organs. In addition, the economic issue of 3D printing can be better addressed by successfully implementing QC, because poor product quality will discourage customers from buying products made by 3D printing. So far, a number of QC programs have been launched to guarantee that products made with 3D printers meet industrial and user requirements. However, QC is still a challenging task for 3D printing because of the following reasons:

- QC relies heavily on the applications of statistical techniques that require a large amount of data and build on experiences. Although 3D printing has been applied to mass production, it is mostly used for prototyping during R&D, for which the volume of production is low. It is also hard to accumulate experiences.
- The quality problems associated with different raw materials, product structures, and/or 3D printing technologies are not the same, making it difficult to apply similar QC techniques. For example, warping is a serious problem when products have elongated or rectangular shapes, but not when products have vertical structures [21].
- The development of 3D printing technologies is still underway. The experience of applying QC techniques to 3D printing is also insufficient. As a result, an absolute rule for choosing suitable QC techniques for a 3D printing process is lacking.

Grieser [32] established a systematic procedure that can be followed to increase the possibility of printing a good-quality 3D object:

Step 1. Set up the 3D printer according to the vendor's instructions;
Step 2. Update the printer's hardware and software regularly;
Step 3. Maintain and calibrate the printer periodically;
Step 4. Clean the print bed before every print;
Step 5. Level the print bed before every print;
Step 6. Adjust the distance between the printhead and the print bed;
Step 7. Print objects of moderate size and complexity only;
Step 8. Choose a filament with sufficient adhesion;
Step 9. Early terminate the printing process if the results of the first few layers are poor;
Step 10. Be patient.

Rengier et al. [38] established a systematic quality assurance (QA) procedure to assess the accuracy and precision of printing human organs. The 3D-printing process was decomposed into three steps: image data acquisition, segmentation and processing, and 3D printing and cleaning. At each step, qualitative inspection and quantitative measurement were performed to validate the printed models.

6.3.1 Product Design

The quality of a 3D-printed object is determined by that of the 3D model. Therefore, the image acquisition step is essential to the quality of a 3D-printed object [39], especially when the object cannot be designed easily using CAD/CAM software. There are various ways to scan an object: optical scanning, ultrasound imaging, computed tomography (CT), etc. Optical scanning captures only the outer shape of an object, while ultrasound imaging and CT can be applied to create the models of internal structures. For this step to be successful, the spatial resolution of the scanning system must be sufficiently high, especially for medical applications [39]. However, there are no absolute guidelines for determining the required spatial resolution. Another way to acquire a 3D model is by converting an existing data file. For example, there have been several software packages for converting a medical image file (usually in the .DIACOM file format) into a 3D-printable model (in the .STL format), such as OsiriX, Invesalius, Meshlab, Meshmixer, and Netfabb Studio Basic [40]. However, many existing 3D databases are heterogeneous, meaning that a dedicated algorithm must be designed for each database.

6.3.2 Process Planning

Segmentation of a 3D model is an important step at the process planning stage [41]. The effectiveness of the segmentation algorithm influences the quality of the 3D-printed object. Common segmentation algorithms include simple region growing, surface/volume rendering, maximal/minimal intensity projection, and multiplanar reformation [39]. However, low-resolution and nonenhanced 3D models require advanced segmentation algorithms [39].

Another task in 3D process planning is to set up the 3D printer by determining the layer height (slice thickness), shell thickness, enabling of retraction, fill density, support type, print speed, and others, which is also a challenging task [21].

Several QC tools, such as expert systems, classification and regression trees (CARTs), and decision trees, can be applied to facilitate the planning of a 3D printing process. Compared with control charts, expert systems may be more effective and practical for 3D printing process planning. After printing various 3D objects, a user accumulates knowledge about how to optimize the setting of the 3D printer for a specific product. This knowledge can be subjectively expressed by the user, or

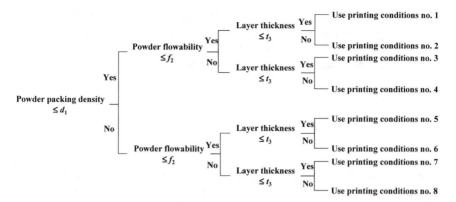

Fig. 6.5 A decision tree

objectively mined using tools such as CARTs [11]. The extracted knowledge can be
stored in a knowledge base on which an expert system can be built, or illustrated with
CARTs or decision trees. An example of such a decision tree is shown in Fig. 6.5.

6.3.3   Cause-and-Effect Analysis

QC techniques can be applied to the planning of a 3D printing process. For example,
the poor quality of a 3D-printed object is attributed to several causes, which can
be summarized with a cause-and-effect diagram, as illustrated in Fig. 6.6. In the
figure, the possible causes of a print failure are classified into the 4 M categories:
man, machine, material, and method. An understanding of products and processes is
necessary, which helps to choose suitable QC techniques for a specific 3D printing
process [10].

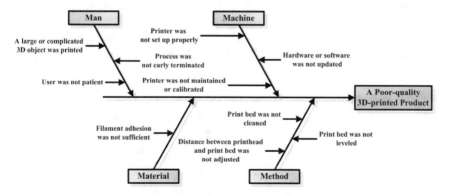

Fig. 6.6 A cause-and-effect diagram (modified from [11])

A cause-and-effect analysis has been conducted by Barclift and Williams [34] to explore the possible causes that led to the low and variable tensile strength and tensile modulus. The possible causes were classified into six categories: man, machine, method, material, measurement, and environment.

6.3.4 DOE and Taguchi Method

DOE, a basic tool in quality engineering, can be applied to optimize the settings of a 3D printer as well as other factors in a 3D-printing process, especially when there are interactions among the factors. However, DOE is rarely accurately and completely applied; instead, the settings of a 3D printer are often determined subjectively, for example, according to the limited experience of the user [42]. A possible reason is the high number of factors to consider. For example, to fabricate scaffolds with 3D printing, the values of at least six major factors (powder packing density, powder flowability, layer thickness, binder drop volume, binder saturation, and powder wettability) must be set to optimize the quality of a printed scaffold, but to discover the appropriate values repetitive and time-consuming experimentation is required [9]. To overcome this problem, Taguchi's orthogonal arrays (or Taguchi's method) can limit the replications of experiments to cover a wide range for each factor [43].

Taguchi method is an experimentation technique for determining the minimum number of experiments that need to be performed by considering the limits of factors and levels [44]. For example, for three factors (e.g., powder packing density, powder flowability, and layer thickness) each with two levels, in theory, 8 (2^3) replications are required to consider all possible combinations to optimize the performance; however, the L4 orthogonal array (see Table 6.2) can be used to achieve the same goal with only four replications.

When each factor has three levels, central composite design (CCD) can be employed to build a quadratic model for the response variable, thereby eliminating the need to conduct a complete three-level factorial experiment [45].

Some examples of applying DOE and Taguchi's method of 3D printing are reviewed as follows. To analyze the effects of three factors on the mechanical properties of a 3D-printed photopolymer part, Barclift and Williams [35] performed a three-factor, two-level full factorial DOE that included eight experiment runs. Hsiao

Table 6.2 L4 orthogonal array

Replication no.	Powder packing density	Powder flowability	Layer thickness
1	Level 1	Level 1	Level 1
2	Level 1	Level 2	Level 2
3	Level 2	Level 1	Level 2
4	Level 2	Level 2	Level 1

[46] designed an experiment of nine runs according to the L9 (3^4) orthogonal array to investigate the effects of four control factors (including nozzle height, printing speed, and UV light exposure time), each with three levels, on two qualities (variation in the droplet diameter and the maximum coverage) of a product fabricated by a photocurable printing system. The results along the two quality dimensions were aggregated using the grey relational analysis method [47]. The values of control factors giving the best grey relational grade were chosen.

6.4   Quality Control in a Three-Dimensional Printing-Based Ubiquitous Manufacturing System

QC is a challenging task for a 3D-based UM system, because it is difficult to guarantee that 3D objects made at different 3D printing facilities achieve consistent quality [48].

With the continuous evolution of 3D printing technologies, the quality of 3D objects printed using newer types of 3D printers is usually better. Therefore, choosing facilities with such 3D printers to be incorporated into a UM system is an essential step to ensure the quality of 3D-printed objects.

In addition, a knowledge hub can be created and maintained by the system administrator, so as to share the knowledge about how to improve the quality of printing a specific type of products.

Subsequently, during a 3D printing process at each 3D printing facility, if the first few layers are unsatisfactory, a common practice is the early termination of the 3D printing process, which avoids the generation of a poor-quality 3D object [32]. In this situation, the UM system administrator has to decide whether to restart the terminated printing process at the same 3D printing facility or not [49]. The reason is that other faster 3D printing facilities may become available when a 3D printing process is terminated at a 3D printing facility.

In addition, when an order is delivered to the customer, his/her feedback on the service and product quality provides valuable information for pushing the 3D printing facilities to improve their operations.

References

1. D.C. Montgomery, *Introduction to Statistical Quality Control* (John Wiley & Sons, New York, 2008)
2. American Society for Quality, Quality Assurance versus Quality Control (2016). http://asq.org/learn-about-quality/quality-assurance-quality-control/overview/overview.html
3. Project Management Institute, Quality Management (2019). https://www.pmi.org/learning/featured-topics/quality
4. Association for Project Management. Introduction to Quality Management (2020). https://www.apm.org.uk/body-of-knowledge/delivery/quality-management/
5. B. Berman, 3-D printing: the new industrial revolution. Bus. Horiz. **55**(2), 155–162 (2012)

6. M. Asadi-Eydivand, M. Solati-Hashjin, A. Farzad, N.A.A. Osman, Effect of technical parameters on porous structure and strength of 3D printed calcium sulfate prototypes. Robot. Comput.-Integr. Manuf. **37**, 57–67 (2016)
7. D.N. Silva, M.G. De Oliveira, E. Meurer, M.I. Meurer, J.V.L. da Silva, A. Santa-Bárbara, Dimensional error in selective laser sintering and 3D-printing of models for craniomaxillary anatomy reconstruction. J. Cranio-Maxillofac. Surg. **36**(8), 443–449 (2008)
8. B.T. Wittbrodt, A.G. Glover, J. Laureto, G.C. Anzalone, D. Oppliger, J.L. Irwin, J.M. Pearce, Life-cycle economic analysis of distributed manufacturing with open-source 3-D printers. Mechatronics **23**(6), 713–726 (2013)
9. S. Bose, S. Vahabzadeh, A. Bandyopadhyay, Bone tissue engineering using 3D printing. Mater. Today **16**(12), 496–504 (2013)
10. J. Norman, R.D. Madurawe, C.M. Moore, M.A. Khan, A. Khairuzzaman, A new chapter in pharmaceutical manufacturing: 3D-printed drug products. Adv. Drug Deliv. Rev. **108**(1), 39–50 (2016)
11. H.C. Wu, T.C.T. Chen, Quality control issues in 3D-printing manufacturing: a review. Rapid Prototyp. J. **24**(3), 607–614 (2018)
12. Y.-C. Lin, T. Chen, A ubiquitous manufacturing network system. Robot. Comput.-Integr. Manuf. **45**, 157–167 (2017)
13. UL, 3D Printing & Additive Manufacturing Equipment Guideline (2015). http://www. ul.com/wp-content/themes/countries/downloads/am/3D-PRINTING-EQUIP-SAFETY-GUI DELINE_EDITION2.pdf
14. T. Orr, ASTM international proposes new 3D printing guidelines for powder bed fusion machines (2014). https://3dprint.com/13367/astm-international-standards/
15. K.V. Wong, A. Hernandez, A review of additive manufacturing. ISRN Mech. Eng. **2012**(208760), 1–10 (2012)
16. R.I. Campbell, M. Martorelli, H.S. Lee, Surface roughness visualisation for rapid prototyping models. CAD Comput. Aided Des. **34**(10), 717–725 (2002)
17. D. Zavorotnitsienko, Understanding 3D printer quality & resolution (2015). http://www.ili os3d.com/en/product-documentation/ilios-documentation-3dprint-quality
18. M. Lanzetta, E. Sachs, Improved surface finish in 3D printing using bimodal powder distribution. Rapid Prototyp. J. **9**(3), 157–166 (2003)
19. R.M. German, Prediction of sintered density for bimodal powder mixtures. Metall. Trans. (Phys. Metall. Mater. Sci.) **23A**(5), 1445–1465 (1992)
20. V. Alfieri, P. Argenio, F. Caiazzo, V. Sergi, Reduction of surface roughness by means of laser processing over additive manufacturing metal parts. Materials **10**, 30 (2017)
21. K.H. Herrmann, C. Gärtner, D. Güllmar, M. Krämer, J.R. Reichenbach, 3D printing of MRI compatible components: why every MRI research group should have a low-budget 3D printer. Med. Eng. Phys. **36**(10), 1373–1380 (2014)
22. J. Shang, Luxexcel and automation & robotics collaborate on a quality control program (2016). https://3dprintingindustry.com/news/luxexcel-automationrobotics-collaborate-qua lity-control-program-95826/
23. N.N. Kumbhar, A.V. Mulay, Post processing methods used to improve surface finish of products which are manufactured by additive manufacturing technologies: a review. J. Inst. Eng. (India): Ser. C **99**(4), 481–487 (2018)
24. J.S. Chohan, R. Singh, Pre and post processing techniques to improve surface characteristics of FDM parts: a state of art review and future applications. Rapid Prototyp. J. **23**(3), 4 (2017)
25. T. Narusawa, J. Shook, *Kaizen Express: Fundamentals for Your Lean Journey* (Lean Enterprise Institute, Boston, MA, 2019)
26. S. Du Preez, A. Johnson, R.F. LeBouf, S.J. Linde, A.B. Stefaniak, J. Du Plessis, Exposures during industrial 3-D printing and post-processing tasks. Rapid Prototyp. J. **24**(5), 865–871 (2018)
27. C. Weller, R. Kleer, F.T. Piller, Economic implications of 3D printing: market structure models in light of additive manufacturing revisited. Int. J. Prod. Econ. **164**, 43–56 (2015)

28. T. Lawton, 3D inspection: how to use a 3D scanner for surface deviation analysis (2019). https://gomeasure3d.com/blog/3d-inspection-how-to-use-a-3d-scanner-for-surface-deviation-analysis/
29. V. Mironov, V. Kasyanov, R.R. Markwald, Organ printing: from bioprinter to organ biofabrication line. Curr. Opin. Biotechnol. 22(5), 667–673 (2011)
30. T. Chen, Y.-C. Wang, An advanced fuzzy approach for modeling the yield improvement of making aircraft parts using 3D printing. Int. J. Adv. Manuf. Technol. 105, 4085–4095 (2019)
31. T. Chen, M.-J.J. Wang, A fuzzy set approach for yield learning modeling in wafer manufacturing. IEEE Trans. Semicond. Manuf. 12(2), 252–258 (1999)
32. F. Grieser, 3D printing quality issues: 10 tricks to avoid them (2015). https://all3dp.com/3d-printing-quality/
33. C. Garrett, How to choose the right 3D printer for you (2016). https://makerhacks.com/choose-3d-printer/
34. N. Hopkinson, P. Dickens, Rapid prototyping for direct manufacture. Rapid Prototyp. J. 7(4), 197–202 (2001)
35. M.W. Barclift, C.B. Williams, Examining variability in the mechanical properties of parts manufactured via polyjet direct 3d printing, in International Solid Freeform Fabrication Symposium (2012), pp. 6–8
36. D. Grossman, Scientists successfully 3D print an organ that mimics lungs (2019). https://www.popularmechanics.com/science/health/a27355578/3d-print-lungs/
37. J. Young, 3D printed aircraft parts and engines could lighten aircrafts by 50% (2015). http://3dprinting.com/aviation/3d-printed-aircraft-parts-could-lighten-aircrafts-by-fifty-percent/
38. F. Rengier, A. Mehndiratta, H. von Tengg-Kobligk, C.M. Zechmann, R. Unterhinninghofen, H.U. Kauczor, F.L. Giesel, 3D printing based on imaging data: review of medical applications. Int. J. Comput. Assist. Radiol. Surg. 5(4), 335–341 (2010)
39. J. Wilding, Make a 3D print from your MRI or CT scan (2020). https://www.printspace3d.com/make-a-3d-print-from-your-mri-or-ct-scan/
40. S. Jacobs, R. Grunert, F.W. Mohr, V. Falk, 3D-Imaging of cardiac structures using 3D heart models for planning in heart surgery: a preliminary study. Interact. CardioVasc. Thorac. Surg. 7(1), 6–9 (2008)
41. J. Herman, 3D slicer settings for beginners–8 things you need to know (2015). https://pinshape.com/blog/3d-slicer-settings-5-things-you-need-to-know-about-3d-printing-software/
42. K. Yang, B. El-Haik, Taguchi's orthogonal array experiment, in Design for Six Sigma: A Roadmap for Product Development (2008), pp. 469–497
43. A. Mohammad, A.M. Al-Ahmari, A. AlFaify, M.K. Mohammed, Effect of melt parameters on density and surface roughness in electron beam melting of gamma titanium aluminide alloy. Rapid Prototyp. J. 23(3), 2 (2017)
44. A. Meena, H.S. Mali, A. Patnaik, S.R. Kumar, Investigation of wear characteristics of dental composites filled with nanohydroxyapatite and mineral trioxide aggregate, in Fundamental Biomaterials: Polymers (2018), pp. 287–305
45. S.-Y. Hsiao, Study of parameter optimization of gradient color process for phto-curable 3D printing system. Ph.D. thesis, Institute of Automation and Control, National Taiwan University of Science and Technology (2015)
46. Y. Kuo, T. Yang, G.W. Huang, The use of grey relational analysis in solving multiple attribute decision-making problems. Comput. Ind. Eng. 55(1), 80–93 (2008)
47. C.K. Chua, C.H. Wong, W.Y. Yeong, Standards, Quality Control, and Measurement Sciences in 3D Printing and Additive Manufacturing (Academic Press, London, 2017)
48. S. Leng, K. McGee, J. Morris, A. Alexander, J. Kuhlmann, T. Vrieze, Cynthia H. McCollough, J. Matsumoto, Anatomic modeling using 3D printing: quality assurance and optimization. 3D Print. Med. 3(1), 6 (2017)
49. T.C.T. Chen, Fuzzy approach for production planning by using a three-dimensional printing-based ubiquitous manufacturing system. AI EDAM 33(4), 458–468 (2019)

Printed in the United States
By Bookmasters